OUT OF THE FIERY FURNACE

Recollections and Meditations
of a Metallurgist

OUT OF THE FIERY FURNACE

Recollections and Meditations
of a Metallurgist

J.A. Charles, BSc, ARSM, MA,
ScD, FREng, FIM

St John's College and
Department of Materials Science and Metallurgy,
University of Cambridge, UK

Book 729
First Published in 2000 by
IOM Communications Ltd
1 Carlton House Terrace
London SW1Y 5DB, UK

ISBN 1 86125 106 8

IOM Communications Ltd
is a wholly-owned subsidiary of
The Institute of Materials

After this text was prepared, attention was drawn to the fact
that the same Biblical quotation had been used for a book by
Robert Raymond, published by Pennsylvania State University
Press in 1986. This described the development of smelting
and metal use in prehistory and there is no
coincidence of content and no plagarism.

Typeset in the UK by
IOM Communications Ltd

Printed and bound in the UK
at Cambridge University Press,
Cambridge, UK

Contents

Introduction

This text is not intended as a full autobiography and will not enter into personal or social matters in my life. What I seek to do is to record, more or less historically, my enjoyment of an unusually varied career as an industrial and academic professional metallurgist. I was most fortunate in being taught in a small group by excellent staff at the Royal School of Mines, at a time when teaching was the primary interest and responsibility of most academics. The 13 enjoyable years in industrial research that followed was at a time before industrial decline had really taken its toll. As an academic at Cambridge, I have had the privilege of interaction with outstanding colleagues, not only in the Department of Materials Science and Metallurgy, but also in St John's College and even more widely in the University as a whole. Teaching bright undergraduates has been a challenge, but also a great pleasure, and the supervision of excellent research students has been little more than gentle guidance, for they have usually quickly outstripped their supervisor!

My early industrial experience was much concerned with furnaces and practical pyrometallurgy in general, whereas after the move to Cambridge, speech, pen and paper dominated the daily round. I had 'come out of the fiery furnace' although not in the sense of deliverance as for Shadrach, Meshach and Abednego, for I had enjoyed my practical pyrometallurgy days. In another sense, also, although most metal products come 'out of a furnace' at some stage of manufacture, pyrometallurgy and furnace technology no longer have a significant presence in the metallurgical content of most materials courses and are themselves 'out'.

It is my aim to bring to notice many people, known and less well known, many sadly deceased, with whom I have had contact and to introduce brief accounts of work, sometimes with them, not always published elsewhere. I shall seek to comment on the many changes that have taken place in metallurgy, its emphasis, industry, research organisations, professional organisation and place in education since my graduation in 1947, and my own assessments of those changes.

Throughout my career I have had continuing care and support from Valerie, my wife, and this volume is dedicated to her.

Imperial College of Science and Technology, 1943–46

ROYAL COLLEGE OF SCIENCE, 1943–44

I left school in Bromley after one year in the sixth form and entered the Intermediate BSc (London University) year at the Royal College of Science where I was taught Physics (mainly by Professor Archer), Chemistry (Dr Emeleus, Dr Anderson, Professor J.M. Heilbron and Dr Sugden as Demonstrator), Pure and Applied Maths (Professor H. Levy) and Scientific German (Dr Sugden). Both Professor Archer and Dr Sugden were killed by 'flying bombs' (V1) later in 1944. The mysteries of mathematics should have been fully mastered under the tutorship of Professor Levy, of high reputation, but although I was greatly impressed by his lectures, held in the old Huxley building next to the Victoria and Albert Museum, lack of practice in much of my subsequent industrial career was not helpful in maintaining that mastery!

Chemistry practicals (group analysis etc.) took place in a huge white-tiled laboratory, lit by pendant lamps with green lampshades from the high ceiling above the old wooden benches. On dark afternoons in the winter, with the lamps on, the thick fumes arising throughout the laboratory were visible as a fog through which the shafts of light were breaking. Around the walls were the fume cupboards, containing Kipps apparati for 'bubbling H_2S'. The smell of the gas was discernible everywhere and by today's regulations we should not have survived. At the walls also sat the Demonstrators; in my part of the laboratory it was Dr Sugden, whose approval had to be sought for any analytical result before proceeding to the next unknown salt.

Before an untimely accidental death in 1932, at the age of 44, my father (Fig. 1) had been a mycologist/biochemist associated with the early work on penicillin (there is a *Penicillium Charlesii* named after him) and my thoughts had turned to taking a Chemistry BSc, and perhaps follow in his footsteps. At the end of the 1944 academic year came the inevitable examinations.

Whilst my overall performance was satisfactory, the competitive 'C' papers in Chemistry were to be my undoing. These papers were set specifically for those who wanted to read Chemistry for the BSc degree from the internal course and from schools throughout the country after taking Highers. I did not appear high enough on the lists to get a Chemistry place. Perhaps I was not cut out to be a

Fig. 1 J.H.V. Charles, circa 1920.

chemist, for within the context of bombs, commuting from Bromley to South Kensington, being a member of the University of London Senior Training Corps (Home Guard), domestic responsibilities etc. I had worked hard and done my best, but what should I do?

THE ROYAL SCHOOL OF MINES, 1944–47

Fortunately for me, the question of my future at Imperial College resolved itself quickly. By then David West (now Professor D.R.F. West), a year ahead of me at school ,whom I had known since the age of eight, had completed a year at Cardiff University reading Metallurgy, and his description of the subject persuaded me that this relatively less popular, historic, 'earthy' field of study, at the Royal School of Mines, combining Applied Sciences in a broad field of industrial activity, with even a hint of 'craft' or 'art', would be much more matched to my capability and interest. I thus applied to the RSM and was interviewed by Professor C.W. Dannatt, whose invitation to 'take a cup of tea' on admission to his office was typical of this old-world gentleman's approach. My overall results in the Intermediate BSc were, in fact, quite good and he was happy to take me onto the course to read Metallurgy.

So, what was the history of the College that I was joining, and of the Metallurgy Department in particular?

History of the Metallurgy Department, Royal School of Mines

Much of what follows relies heavily on an internal document, prepared by Dr M.S. Fisher in 1963, and on a booklet issued on the occasion of the RSM's centenary in 1951. It is, in fact, the oldest metallurgy department in the UK and has its origins in the Government School of Mines and Science Applied to the Arts, which opened in Jermyn Street, London in 1851. Several, later, metallurgy departments in other universities were to be founded by its Associates. Emphasis was to be on subjects directly related to mining and the treatment was to be essentially practical and technical. A formidable staff, mostly Fellows of the Royal Society, included Sir Henry de la Beche (mining) and John Percy, the latter justly thought of as the 'father' of metallurgy an applied science. There were about 60 students, but only 12 of these were taking the full-time two-year course. With attempts to improve recruitment, the name was changed to the Royal School of Mines in 1862 and the course extended to three years, covering mining, mining mechanics and surveying, metallurgy, practical geology, mineralogy, palaeontology and mechanical drawing. The qualification arising from the course was Associateship of the Royal School of Mines, but in spite of the genuinely high standing of the award the numbers of full-time students remained low and there were doubts about the survival of the school. The majority view held by the staff then, and it is a cautionary tale, was that there were too many lectures and too little practical instruction, for which there was inadequate space. In 1872 the authorities recommended a move to what became known as the Huxley Building in Exhibition Road. In this, to my mind, unattractive building, I attended first year courses (1943–44) in applied mathematics and statistics. I recall that the Senior Training Corps (Home Guard) had some basement space.

The Geology Department moved into the Huxley building in 1879, followed quickly by Metallurgy, the latter occupying the basements. The Head of the Metallurgy Department at Jermyn Street had been Professor Percy, who opposed the move, fearing that his younger colleague Huxley was aiming to start a college of general science, into which metallurgy would be subsumed, losing its identity. So opposed was Percy, when the move happened he resigned his chair. Huxley did, in fact, form his new college of science, comprising biology, chemistry, mathematics, mechanics and physics, which soon became the Royal College of Science. Percy need not have worried, to the extent of resignation, since the RSM was not incorporated into Huxley's new college; although it was reduced to two departments, Mining and Metallurgy, there was a link to the Royal College of Science through the Geology Department, the latter eventually to be absorbed by the RCS. All housed in the same building, there was, of course, co-operation between the two Colleges in the teaching of courses such as mathematics, mechanics and physics, just as many years later subsidiary subject teaching in the Metallurgy course was provided by the RCS and City and Guilds College.

Roberts-Austen, Chemist and Assayer of the Royal Mint, succeeded Percy, although retaining his Mint post. The RSM flourished, but mainly again in relation to special courses providing instruction for company employees, with the number of Associates still at about only 10 per year. With the growth of the other subjects in the Huxley Building, there was serious overcrowding and in 1905 very grand buildings were provided for Chemistry and Physics in Imperial Institute Road. Very familiar to me in my first year (1943–44), I witnessed their demolition in later years with great sadness. Metallurgy was left behind in the Huxley Building. Meanwhile, another new college, known initially as the Central Technical College, had been built by the City of London Institute for the Advancement of Technical Education, providing instruction in applied science and technology for students of civil, mechanical and electrical engineering and 'technical chemistry' (chemical engineering); this became the City and Guilds College.

There were now three independent College teaching aspects of science and technology in South Kensington, and as can be imagined there was a great deal of overlap in administration and in the courses. In 1907, a Royal Charter incorporated the three into the Imperial College of Science and Technology. Meanwhile, Metallurgy was still in the Huxley basement. Roberts-Austen had greatly widened the course, particularly through the introduction of the newly-developed techniques in metallography, but was forced to retire through ill-health in 1902. He was to be succeeded by Professor Gowland and, in turn, by Professor Carlyle. Whilst Roberts-Austen had introduced more physical metallurgy, the 1912 Associate syllabus was still heavily biased to mining, extractive metallurgy and assaying as satisfying the current vocational opportunities, although, interestingly, the research school was equally concerned with physical metallurgy topics. In 1913, Carlyle was succeeded by a genuine physical metallurgist, H.C.H. Carpenter from Manchester University. Not only were the Associateship courses to be completely reorganised by Harold Carpenter after the First World War, but from 1910 onwards there was a steady relocation of mining, metallurgy and geology to fine new buildings in Prince Consort Road. The relocation started with the Bessemer Laboratory, to which the Bessemer Memorial Committee had donated £10,000. The rest of the building (designed by Sir Aston Webb in Portland Stone) was finished onto the Prince Consort Road frontage in 1915. Thankfully, although the Bessemer Laboratory has almost gone, the main building remains and has not suffered the same demolition fate as so many other fine old buildings on the Imperial College site; the facade at least is listed. The new 1915 building was very commodious for both students and research workers alike, fortunately so since it allowed for considerable expansion in numbers for many years later.

Carpenter was keenly interested in all aspects of physical metallurgy, as reflected in the two-volume work *Metals* (written with one of the lecturers in his department, J.M. Robertson, and published in 1939), but particularly in alloy constitution and crystallisation, as in his work with Miss C.F. Elam (later Mrs Tipper) on recrystallisation and grain growth. Hume-Rothery was another very gifted member of his research school at the RSM. William Hume-Rothery, eventually became Head of the Department of Metallurgy at Oxford University and Mrs Tipper a Reader in the Department of Engineering at Cambridge. Significantly, there was no research activity in the field of non-ferrous extraction metallurgy, although this was where much of its teaching effort and reputation still lay. This was a situation not greatly altered until the Nuffield Research Group in Extraction Metallurgy was formed under Dr F.D. Richardson in 1949. Although Professor Sir Harold Carpenter gradually introduced more physical metallurgy topics into the course to match industrial need, there was still an overemphasis on assaying and chemical analysis, even in 1944 when I began, with steadily less likelihood of such skills being involved in the jobs available. Whilst, subsequently, I found subsidiary geology, mineralogy, mineral dressing, engineering drawing, applied maths etc. to be useful in my career, I must admit to not having used practical chemical analysis skills at all.

At the start of the Second World War, the third and fourth years in the Metallurgy Department were transferred to Swansea University with the first and second year courses remaining behind, largely provided for by the RCS and City and Guilds with such metallurgy teaching as there was at that level being provided by Dr S.W. Smith, honorary lecturer and later Assistant Director of the Department. Professor Carpenter died in 1940 and his responsibilities were taken on by Assistant Professor C.W. Dannatt. The Department returned to South Kensington at the end of the 1942–43 accademic year.

Professor C.W. Dannatt

C.W. Dannatt (Fig. 2) was a former student of the Royal School of Mines, entering in 1910 and graduating in metallurgy in 1914. He then joined the Queens Westminster Rifles, was commissioned in 1915 and served in France, Salonika and Palestine. After 1918, he held two technical posts, one in Egypt and the other in Trinidad, and then returned to the RSM in 1923 as a Demonstrator in Metallurgy, rising to Assistant Professor and Reader in 1937. Always a modest and kindly man, students were largely unaware that he was an accomplished pianist and an outstanding tennis player.

During the 1943–46 period, Professor Dannatt, as he was by then, made major changes to the course, introducing a further emphasis on physical metallurgy at a more advanced level and on ferrous metallurgy generally with the appointment of several new members of staff. The new course, from which I benefited greatly in my final two years, remained essentially unchanged until his retirement in 1957. Steadily, the amount of time spent on assaying and chemical analysis had

Fig. 2 Professor C.W. Dannatt.

been reduced, although non-ferrous extraction metallurgy still retained great importance. In this context, it has to be remembered that the majority of the graduates in 1947 would still go into non-ferrous metal extraction posts. That Carpenter, a physical metallurgist, had not succeeded in making such changes, whereas Dannatt, a skilled analyst, extraction metallurgist and ardent tradition-alist, was prepared and able to do so, is a remarkable fact. Certainly job opportu-nities were beginning to move away from the non-ferrous extraction side and he recognised the growing importance of physical metallurgy in the changing op-portunities for graduates. He did not deserve the criticism made of him later for refusing to change back to more chemical aspects of the subject under pressure from the Nuffield Group; for it was he that had obtained the starting grant of £5000 per year for five years from the Nuffield Foundation, with the backing of the Institution of Mining and Metallurgy, to establish a Fellowship for physico-chemical extractive metallurgy research in the Department. Dr F.D. Richardson was appointed to the Fellowship, and then to a new Chair of Extraction Metal-lurgy when Professor Dannatt retired in 1957. 'FDR' retired through ill-health in 1976, although he remained a Senior Research Fellow until his death in 1983. My relationship with him commenced during my British Oxygen Company days (1950–60) when attending the very valuable British Iron and Steel Research Association conferences, where his mentor from his Navy days, Sir Charles

Goodeve, was also often present. An important associate of FDR was J.H.E. Jeffes. Jim gave him great support over many years and was then, and still is, approachable and amiably helpful when his advice is sought on any physico-chemical matters.

Process engineering aspects of research at the RSM developed into the John Percy Research Group, led by A.V. Bradshaw, again supported by the Nuffield Foundation. Tony later emigrated to Australia and the CSIRO. The first John Percy Research Fellow was A.W.D. Hills (David), a Cambridge graduate who I came to know very well in later years when he had become Head of the Department of Metallurgy at Sheffield Polytechnic (later Sheffield Hallam University), when I was frequently acting as External Examiner for different courses and research theses. Many happy evenings were spent with David and his wife Jane.

First Year, 1944–45

That then, was the history of the Department to which I had come. There would eventually be 16 students, with the rest coming straight from school bar four: myself, Peter Harding, a repatriated Spitfire pilot released from a prisoner-of-war camp who had started the course before the war, and two Polish Army Officers, J. Fiegel and F.S. Paschek, the former a Polish 'VC' from Anzio, who joined the course later in the 1945–46 academic year. Those who had come from school had gained State Scholarships in Higher School Certificate examinations (the equivalent of A-level today).

The Royal School of Mines was, in fact, very much smaller than either of the other two constituent colleges of Imperial, City and Guilds College and the RCS, with about only 60 students taking the various courses as compared with about 1400 at IC altogether. Geology had remained part of the RCS, a situation not changed until the 1960s. Its small size was not reflected, however, in its impact on the sport and social life at IC. As a close-knit community, there was an appreciation that participation and effort mattered and we were well able to 'keep our end up' in both social and sporting events in spite of our small size.

Incredibly, many of the students in my entry year were eventually to find themselves working at the Imperial Smelting Corporation at Avonmouth, mostly after National Service during 1947–49, and I was to be associated with some of them over the 25 years or so that I acted as a consultant to the company: Alan Self, Colin Harris, 'Phil' Gray, 'Bluey' Mechem, Alan Eckersall, 'Dick' Healy. Of the others, Peter Harding went to work at the Enthoven, Rotherhithe, lead smelter, one of the Williams twins ('Wis' and 'Was') went to Gillette and the other, I think, to London Scandinavian. 'Willy' Hall went out to the Copper Belt in East Africa, as did J.C. Loretto. Bill Pollard initially joined British Aluminium at Banbury (now Alcan) before emigrating to Canada and working in the Government Bureau of Mines Research Laboratory in Ottawa for the rest of his career.

He and Phil Gray, along with myself, were both also members of the University of London Choir, and my particular friends in the group. B.T. Houlden worked very hard and duly obtained a First and stayed on to do research for a PhD. I have not had any news of him since then. With the exception of J.C. Loretto, we are all seen gathered on the steps of the RSM in Fig. 3.

The very encouraging introduction to the RSM that I had in that first interview with Professor Dannatt was the start of three very enjoyable years, in spite of the background of war persisting in the early part. Without a doubt I had found a subject which suited me and that has continued to fascinate me over a long professional career. In the RSM there was an approachability and friendliness on the part of the staff, which I subsequently found to be very much a characteristic of the metallurgical profession.

Whilst commuting daily from Bromley to South Kensington (via Bromley South and Victoria) had obvious disadvantages, it also had its lighter side in that many youngsters that I knew, male and female, were also making a similar journey to London and friendships were made or developed in doing so; in my case I found the girl who was to be my wife! In any case, at that time there were only a few hostel places available and these were for people who could not live at home. Journeys could be very difficult; Southern Railway rolling stock was old, with minimum wartime maintenance and breakdowns were not infrequent. These were often associated with power failure, which in the bombing and flying bomb and rocket periods was sometimes associated with damage to trackside installations. For a breakdown to occur in Penge tunnel, with up to 20 people packed in a compartment with seating for 10, the rest standing, and all in the dark, is an experience not readily forgotten! Fog was another major cause for delay when trains could be an hour or two late. Whilst atmospheric pollution is still a major problem in relation to health, it does not obscure vision in the same way as then when coal was the major fuel in use. One remembers the rush to the heavy leather straps to raise windows when steam trains went into tunnels.

Once started at the RSM, I was in for one or two shocks. My first RSM Union meeting, where attendance was obligatory for fear of dire consequences if one did not, introduced me to the more sordid features of young adult male group culture, perhaps more than I had experienced even at Army camps the year before. The lecture room used was full of bodies and smoke, with most men pulling on pipes or cigarettes and reclining with feet on the desk in front. Whilst the formal matters of reading Minutes, making reports etc. was carried on at the front by the Secretary, President, Treasurer etc., the 'Official Pornographer' was showing a series of sexually explicit photographic and cartoon slides to ribald comments from the audience. It was the job of the OP to add further slides to the collection in his year of office and also another verse to the poem "The Exploits of Eskimo Nell", written in a specially provided brass-bound book. When

Fig. 3 The final year Metallurgy class, RSM 1947.

women first entered the RSM, well after I had left, this form of meeting was clearly not acceptable and, after a rather hopeless attempt to exclude females, these sexist and potentially embarrassing aspects were removed; I believe the meetings now follow more conventional lines. In a way, the efforts that had to be made to keep the RSM in contention with the much larger RCS and City and Guilds colleges, within the overall Imperial College, produced a more united, socially aggressive and close-knit community, and to an extent the form of the meetings followed from this.

The second shock was the Freshers Dinner, where one was hosted by a second or third year undergraduate 'to look after you', after which all freshmen were required to undertake an initiation of being required to stand on the bar and drink a 'yard' of ale. Not being in the habit of drinking over two pints of liquid all at once, shooting down the stem at substantial velocity, it was quite an ordeal and certainly strained the digestive and excretory system subsequently. Once 'in', however, one was able to establish one's own position in relation to the social activities of the group. In fact, the metallurgists were, in the main, the most restrained set in the RSM, certainly in relation to the mining engineers. In any case, as regards drinking financial restraints ensured strict moderation with the family budget under strain. Thus it was, that, in the subsequent years to 1947, I was frequently the one to assist various friends on the route to South Kensington station, on the way home to catch a late train!

The Metallurgy course at the RSM was an extremely good one. As recalled earlier, it retained some of the pre-War emphasis on extraction metallurgy and assaying, but it was by then much diluted with modern physical chemistry and physical metallurgy. It was an extremely broad course as indicated by the major examination papers that were taken, with the results accumulating to the final grade of degree. The courses that I undertook were:

1st year (1944–45): Applied Mathematics I and II; Engineering Drawing and Design; Assaying; Geology; Electrical Engineering; Workshop Practice

2nd year (1945–46): Extraction Metallurgy (Ferrous); Mineral Dressing; Refractory Materials; Metallography and Pyrometry; Metallurgy (General and Non-Ferrous)

3rd Year (1946–47): Extraction Metallurgy (Non-Ferrous); Physical Metallurgy; Mechanical Treatment and Testing; Metallurgical Analysis.

There was a very strong practical content in the course overall, reflected in four-day practical examinations in each of assaying, metallurgical analysis and metallography. The RSM was particularly well equipped in the mineral dressing area and for assaying and analysis, but modern equipment for metallography and physical metallurgy was also being brought in during 1945–47. The RSM was the centre for BSc (London) external practical examinations for institutions, such as Battersea Polytechnic and the Sir John Cass Institute, reflecting the facilities available.

I greatly enjoyed the subsidiary Geology course, taught by Professor H.H. Read, Professor David Williams and Dr Gilbert Wilson. George Sweeting was the Demonstrator and he made the recognition of minerals by hand and with charcoal block, candle flame and blowpipe good fun. For the first time we were introduced to thin section microscopy and the subtleties of polarised light, calling on our best artistic skills in representation (I still have the coloured drawings). We were joined in this course by one of the few girls at Imperial, Rosemary Lacey, extremely attractive and already associated with one Peter Baxendell, an oil technologist a year ahead of us, whom she later married. He was later to become Chairman of Shell, and then Chairman of Hawker Siddeley Electric Export Ltd.

George Sweeting was to play a central part in taking us on the Geology field trip to Ashover, in Derbyshire. It was my first real trip away from family associations and environs in the south east and East Anglia, and I was captivated by the fell countryside. The centre of Ashover itself has many old stone houses with that mellowed look that only centuries can give and was then possibly the most attractive village in the valleys and dales of Derbyshire. High, steeply pitched roofs suggest an earlier use of thatch and in so many ways it was a counterpart to better-known villages in the Cotswolds. Ashover is in an ideal area for a field trip giving many opportunities for mapping, mineral/fossil collecting etc. In a glaciated

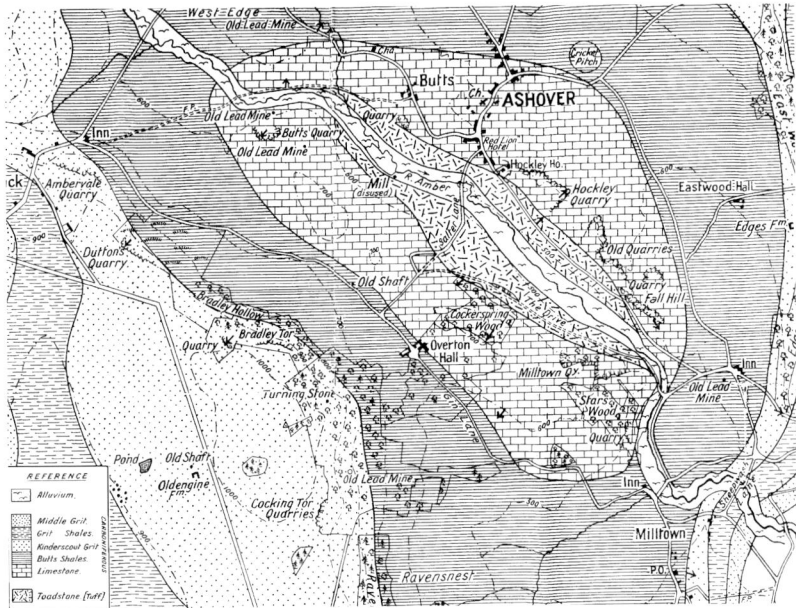

Fig. 4   The geology of Ashover, Debyshire,(George Sweeting, Proc. Geological Association of London, 1946, p.22).

valley between two millstone grit ridges is a carboniferous limestone bed containing quarries, shale outcrops and a volcanic 'tuff' inlier (Fig. 4). The quarries yielded fossils of all sorts, particularly corals. Certain areas contained lead mining shafts, where the spoil heaps were being worked over for the previously rejected fluorspar and where numerous copper, zinc and lead minerals could still be found. The fluorspar was required as a flux in steelmaking.

We students were billeted around the village. I can only recall that my landlady gave me excellent breakfasts and that rationing clearly had less impact in the country as far as bacon and eggs and bread and butter were concerned. The headquarters for the week was the Red Lion, where a room was set aside for us in the evening to complete geological mapping from the days walking and much beer was consumed. Packed lunches were taken at some vantage point such as the 'Mines Stone', high above the Ashover valley on the millstone grit, or at the Miners Arms, where time was taken for beer and the Skittle alley. It was, in fact, one of the most enjoyable times I had experienced to that point. My appreciation of the whole Geology course was reflected in the high First class marks achieved at the end of the year, and I was approached by Professor David Williams to see if I wanted to change to Mining Geology instead of Metallurgy for my degree. It was not particularly difficult to say 'thank you, but no'. I liked everything that I saw ahead of me in the Metallurgy course.

the flow over the strakes were passed to a Pachuca air-lift agitator with the addition of sodium cyanide for the final recovery. Control of the leach was through regular sampling and testing over, I think, a 24 hour period. Certainly there was a night shift of four, where the procedure of moving over the open-top Pachuca containing cyanide and lowering a jam jar on a string for the sample, fortified by the odd beer, would also certainly be forbidden by today's dictates for safe laboratory practice!

A particular course taken at this time was Metallurgical Calculations, which consisted mainly of generating charge balances in smelting and refining, calculating required slag-making additions etc. The calculations, using log tables or slide rules were often long and complicated and no doubt computer programmes exist today to carry out such tasks. I certainly doubt if any material science students today would be able to cope.

During 1945–46 and further in 1946–47 a lot of course time was put into metallography and for me the examination and interpretation of metallic microstructures was of particular fascination, and so it has always remained. My microscope bench was immediately under a window looking out onto Prince Consort Road and an unusual hazard were the flashes of illumination from the mirrors of office girls occupying one of the sun-lit south-facing upper storeys of the building opposite, who in the summer, took their breaks on the balconies outside their windows. The class responded with suitably robust posters on our windows until the authorities intervened.

Stewarts and Lloyds, Summer 1946

Bill Pollard and I both elected to do our eight week spell of vacation work during the summer of 1946 at the Stewarts and Lloyds iron and steel works in Corby, Northants. Built in 1935 by Brassert Engineering to exploit the local Jurassic ironstone, it was still regarded a modern plant in 1946 with the latest blast furnace design and basic steelmaking converters. One of the famous names of the ferrous industry, Tom Colclough, immortalised by one of the Institute of Materials' medals, had been involved in the building and commissioning of the plant.

It was a good example of a then modern, integrated steelworks and was unique, not only because of the processes involved, but also because of the extremely low grade ore which was used to give, at the finish, a cheap steel. Low transport costs (local ore), large scale manufacture by the standard of the time, every available by-product recovered and sold, economy in coke consumption and complete integration, made the exploitation of the ironstone at the time a profitable enterprise in the face of competition from even plants based on imported foreign ore. Although the local iron ore was low grade and phosphoric, it was largely self-fluxing and further dilution of the blast furnace charge by fluxes other than for charge adjustment was not necessary, reducing the coke requirement. The consequent desulphurisation of the iron that was required was subsequently achieved with soda ash in ladles.

The competitiveness of the plant was to change in later years, however. This was principally because very large dedicated bulk ore carriers bringing very rich ore from abroad to coastal plants was at very low cost and also meant a much lower slag volume per ton of iron and a further reduced coke consumption. Now that the plant now no longer exists and Corby industry has changed in character, a brief description of the plant and the practice followed at that time may be of some historical interest.

The mined rock could be classified as stratified ironstone of the Jurassic system, the Northamptonshire ironstone being one of the three main beds (the others being Marlstone and Frodingham). The majority of the workings were opencast, although three underground mines were still in operation. At that time the ore was assaying at 28-34%Fe, with 7-20%Si, 4-12%Al_2O_3, 1-25%CaO, 8-22%H_2O, with reserves of 67,400 acres and an average yield of 2000–2500 ton/ft/acre. A mechanical excavator took off the top lift which was moved to the area already excavated for use in reclamation. The main overburden or bottom-lift was then stripped off with a 650 ton drag line excavator making a 50 ft cut, the ironstone floor was broken up by blasting and then dug out. The quarries were worked and drawn upon to give as balanced and constant an ore as possible at the blast furnace plant, aiming for a self-fluxing system at about 30%Fe. Care was taken in relation to the regions of the ironstone containing pyrite or gypsum, with control through the depth of cut being taken from the ironstone bed. The bed was normally richer in iron towards the bottom.

Underground working, as at Church Mine, Islip, was on the pit and stall system, entered at each end from the bottoms of old quarries. The nature of the overburden meant that very little propping was needed and allowed for lofty and wide workings with natural ventilation. The floors of both quarries and mines was the top Lias formation, known by the men as 'green' bastard rock.

The Islip Blast Furnaces
Originally, all the ore from the Church Mine went to the old blast furnaces at nearby Islip, but in 1946 all of the approximately 2000 tons of ore extracted per week joined the other sources feeding the furnaces at Corby. Although my interest in historical metallurgy, and archaeometallurgy in general, had not developed at that time, it was an incredible experience to be able to roam around the Islip plant, which had closed in 1942 with the redeployment of manpower to Corby. The blast furnaces were erected in 1872, there being originally four furnaces and eight Cowper two-pass stoves. Only three ever worked together and one had already been dismantled. They were 70 ft high with a hearth diameter of 10 ft. The double bell mechanism was hand charged using iron wheelbarrows (almost

too heavy to move when empty!), brought up by lift from the ore bays. Output was about 150 tons/day for each working furnace, cast into 1 cwt pigs. Motive power for the lifts was steam, as for the reciprocating engines driving the piston/cylinder pumps providing blast. Calcining of ore was carried out in the brick ore bays themselves when the furnaces were in operation. An interesting by-product from Islip was ganister. Considerable quantities occurred on top of the iron-stone and historic waste dumps were being worked to produce a saleable refractory product. Although all was now quiet, almost ghostly, it was not difficult to imagine the physical conditions under which men had once worked, and to marvel at their toughness.

As previously indicated, all the ore handling practice at Corby was designed to give a balanced ore of constant composition, essential for good blast furnace operation. Trucks containing ore were assembled in series and fed to the crushing plant according to a prearranged scheme. After sizing, the fines went to sinter strands and the lumps, 3–4 inch in size, to the furnaces direct. In spite of all the care taken, however, fluctuations in charge analysis did occur, and in 1946 an ore bedding plant was commencing operation where a large layered bed was sliced through for charging.

The Corby Plant
In 1946 there were four blast furnaces (three 20 ft hearths and one 18 ft, with heights of 95–85 ft), each equipped with three Brassert stoves. They were operated on an acid slag of lime/silica ratio of approximately unity. Sulphur was completely disregarded, the iron being desulphurised in transfer ladles. The low slag ratio gave a low melting point, and thus the furnace could be run at a lower temperature with a low slag volume and thus maximum elimination of silicon to the slag. With this practice also, the slag, being low in lime, did not weather and was suitable for use in the manufacture of Tarmac after crushing and dressing. Some phosphorus-rich slag recycling was carried out, returning phosphorus to the metal and thus adequate exothermicity during conversion and a saleable converter slag for use as a fertiliser ($+16\%P_2O_5$), another example of the emphasis on maximising revenue from by-products. The pig iron made, analysed in the transfer ladles after some sulphur removal with soda ash, was 0.65%Si, 2.0%P, 0.6%Mn and 0.1%S. Tapping was every six hours, thus with all four furnaces running there was a delivery of hot metal to the mixers at the converter plant every 1 hour, with each tap being of about 100 tons in two 50 ton transfer ladles.

The blast furnace stoves on the Corby plant were Brassert patent zone-filled units, three to each furnace changing at intervals of four hours from 'on gas' to 'soaking' to 'on blast'. Cleaned blast furnace gas was used for heating the stoves. From the furnace downcomer, the crude gas went through dust catchers (Howden Vortex), Brassert washing towers, disintegrators, Lodge Cottrell washers, and

Lodge Cottrell electrostatic precipitators. Fifteen per cent of the cleaned gas was employed directly on the stoves, the rest going to a gas holder. I remember at the time being particularly struck by the major feature of water supply and quality control for the scrubbing systems, steam boilers and for general water cooling purposes. The various water sources (mainly company reservoirs) had different permanent and temporary hardnesses, requiring careful control in the softening plant, which had a capacity of about 170,000 gallons.

Four steam turbines working at 3400–3800 rpm drove turboblowers with output 42,500–50,000 ft³/min each, providing the air blast for the furnaces. The steam for the turbines was generated by five Babcox and Wilcox multitubular boilers operated on coke oven gas/blast furnace gas mixture, generating a maximum steam flow of 50,000 lb of steam per hour at 350 psi. Sodium metaphosphate was added to the feed to the boilers for final hardness removal, with removal of any residual sludge from the blow-down water. All this could perhaps be considered small beer today, but at the time it was extemely impressive to me.

The most expensive component of the blast furnace charge was, even then, the coke, and considerable attention was paid to the control of the coke oven operations to give a consistent coke of required properties. At that time the best blend of coal for the ovens was apparently 70% of Silkstone A+B (e.g. Maltby, Denaby, S. Kirkley, Chatterley, Bolsover) and 15% of Durham (Pelaw, Holmside) and 15% Welsh or Silkstone C (Selstone, Riddings, Alfreton, Ackton Hall). The inclusion of this information is merely to underline the number of collieries working at that time, giving considerable flexibility in the production of a blend to give the desired properties, but requiring very careful marshalling and delivery to the crushing and storage bins! At the time the coking plant was the biggest in Great Britain and the second largest in Europe.

In hindsight, one recognises the energy wasted during cooling after discharge, with clear memories of the discomfort of that side of the ovens, a feature now often dealt with in modern practice. Cleaning the coke oven gas also produced a considerable number of saleable by-products such as tar, ammonia, benzol and toluene. Although Bill Pollard and I recognised the importance of this by-product recovery plant, we did not enjoy the detailed investigation. Perhaps it was because we were not chemists, but more likely we were put off by the smells!

The heart of the Corby plant was, for us, the converter shop. Three 1000 ton inert mixers (usually two in operation), fired with a 4:1 mixture of blast furnace and coke oven gas, supplied five 25 ton basic Bessemer converters with hot metal as required. A straight line arrangement of the mixers and converters facilitated the crane transfer of hot metal from one to the other (Fig. 6). A certain amount of slag that had gathered in the transfer ladles was rabbled off prior to charging the mixers. The mixers poured into ladles containing some soda ash for further sulphur control which then transferred the contents to the converters. After the

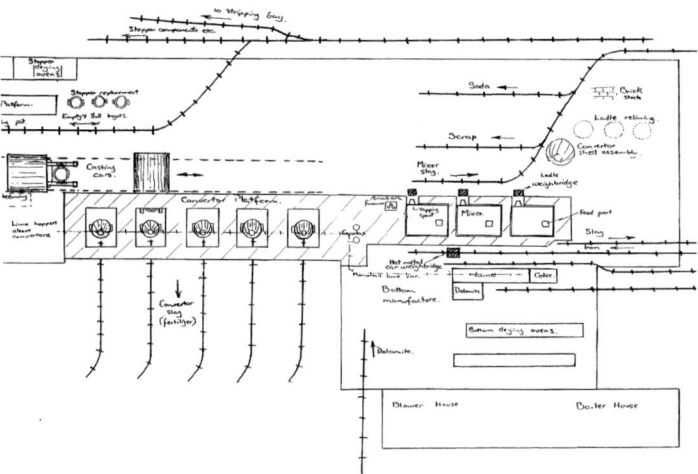

Fig. 6 General layout of converter shop, Stewarts and Lloyds, Corby, 1946 (from JAC report).

blow, the steel was poured into a 25 ton teeming ladle, supported by one of two casting cars, the arms of the casting car being able to swing the ladle over the teeming pit and the ingot mounds.

A particularly impressive feature of the converter operation, apart from the usual pyrotechnics associated with tilting up, was the judgement of steel condition by the 'blower' (the other important member of the team on a converter was styled the steersman, responsible for tilting etc.) At the start of a blow operation followed the appearance of the flame, initially non-luminous, bluish in colour and with many 'sparks' as most of the silicon and manganese was removed over a period of about 2 minutes. Then, as the carbon removal got under way the flame became yellow and luminous, lengthening to about 30 ft, for a period lasting about 8 minutes. The end of this 'foreblow' was signalled by the flame shortening to only about 2 ft long and becoming non-luminous. For phosphorus removal, of course, there was then an 'afterblow' of 3–4 minutes during which copious brown fumes (iron oxide) arose from the converter mouth. Progress was judged by the appearance of spoon samples observed through a blue glass with each flame appearing to be individual to the particular 'blower' but seemingly varying in intensity of colour from one to another. Sampling was carried out at 20–30 second intervals and judgement was made according to the surface appearance in terms of colour and distribution of bubbles. An even distribution of fairly small bubbles was desired, presumably reflecting a satisfactory oxidation state in the evolution of carbon monoxide. In special circumstances, spoon samples were subjected to fracture tests, in particular when the mixers were short of metal and the buffer capacity was not ironing out greater variations in composition than normal, with fracture assessment made on the basis of lustre and grain size. It was later in my career that I accepted a challenge from the first hand on an open hearth to assess the carbon content of a fractured spoon sample. I felt that if I failed it would merely reinforce the challenger's feeling of technical superiority in a different field and do me no harm and if, by sheer luck, I succeeded, my own stock would be raised. My luck held!

The appearance of the slag on the steel surface was also taken as a guide to the condition of the bath. A clean, smooth slag indicated satisfactory condition. A wrinkled surface indicated a low temperature. Scum, i.e. floating dross, resulted from too high a lime content. Too high an FeO content or a high temperature gave a frothy, bubbly slag. Where required, increase in slag fluidity was obtained by the addition of sand to the predominantly basic slag required for the phosphorus removal.

Carbon and manganese content adjustment was made at the end of the afterblow with additions of ferromanganese or spiegel, and deoxidation in the ladle by ferromanganese and aluminium. Spiegel raised the carbon content with a smaller increase in manganese and was preferred unless a special high-manganese grade was required, since ingot mould attack was lower and rimming more successful. Where the steel was to be subsequently cold rolled, lower phosphorus and sulphur contents were specified, requiring a double slag blow (a further charge of lime made after the first slag removal and the blowing continued for a further 30 seconds). Sulphur was lowered by the addition of further soda ash. Typical analyses were:

	C%	Mn%	S%	P%
Single slag blow	0.04	0.37	0.039	0.045
Double slag blow	0.04	0.37	0.027	0.024

The above notes underline the great changes that have taken place in converter control. The only samples taken for analysis then were from the mixer from time to time, from the ladle from each blow as a record of achievement and from ladles during teeming. Today, with a range of sensors available for determining the bath condition as refining takes place and with computerised reaction to changes, much less is left to the skill of the blower. Whilst acknowledging the fact that everything was done to ensure consistency of converter supplies as the basis for control, from experience, the ability of the operators was awesome.

Two main types of ingots were teemed: 'fully rimmed' and 'plugged'. In the latter the rimming action was not allowed to proceed to completion, using a bottle shaped mould where a plug or cap could be dropped into the neck. This gave a partially rimmed structure where impurity segregates were moved further away from the sides of the ingot at the top, leaving a clean rim and reducing scrap, particularly where the Mannesmann push-bench was being employed for subsequent tube production, which was the main product from the Corby plant. This characteristic was to have some significance when, many years later, I became involved in studies of ingot structures.

Steel was also made at Corby in two basic 25 ton electric arc furnaces installed in an extension of the converter shop, dealing mainly with internal scrap from the tube mills and using fairly standard arc furnace practice, with both voltage and current control. A short circuiting device, by means of nichrome wire mesh inserted in the furnace hearth, acted as a precaution against breaking electrodes through driving into the scrap during the irregular conditions at the beginning of the melt. The electrical engineering course that we had attended at City and Guilds during my first year at the RSM came in useful then when trying to work out the furnace control systems. With the ability to work with a fully fluid basic, reducing slag, high quality steels could be produced. During the World War II these were often special alloy steels for the armaments industry, but in 1946 straight carbon steels, of controlled lower phosphorus and sulphur, were made for high quality seamless tube production.

Ingot production from both converter and electric arc plant went on to the 40 inch two-high reversing blooming mill and then to a 32 inch reversing mill, finishing to the required size for the next series of operations, either to $6^5/_8$ inch square for the push-bench tube production or slab for the 'skelp' or strip to be used in one of the three Fretz-Moon continuous welding mills according to size. Emerging from a heating furnace at near-welding temperature the continuous strip, flash-butt welded from coils, was locally heated at the edges by flat nozzles delivering an air jet prior to curved rolls bringing them together with a further air jet in a pressure weld. A flying saw followed. As with continuous strip mills, the product tube emerged at high speed and if for any reason cutting to lengths failed, a fearsome tangle or 'cobble' could result.

The plant was operated almost entirely by Scots, many of whom had walked down from Lanarkshire in the late 1930s on the promise of work, and in the later generations of Corby folk the Scottish accent can still be detected. We were under the care of the chief metallurgist, Dr John Glen, who was extremely kind. A programme was devised whereby we witnessed every phase of the operations. My vacation work report covered all the technical information gleaned, but did not reveal some of the more exciting moments. To a considerable extent we were left to our own devices once arrangements had been made to visit a particular plant. This was to lead to dangerous situations, impossible to envisage today, on at least two occasions. In one, we took the lift to the top of the blast furnace stoves to gain access to the charging bell platforms of the furnaces, via narrow steel mesh catwalks, in order to determine the bell rotation sequences being employed for even charge distribution. Unexpectedly, although in the open air, the leakage of carbon monoxide-rich gas from the bell seals was sufficient to cause dizziness some 85 ft up before we staggered away to the catwalk, into fresh air. On another occasion we decided to hitch a lift on the overhead conveyor belt which was moving crushed coal from the coal blending plant to the coke ovens. It became apparent that we were in danger of being precipitated into the hopper with the coal, but fortunately managed to jump clear!

During the last few weeks I was given a 'real' job to do, as metallurgical observer on the teeming platform in the Bessemer plant. Provided with a large clipboard and printed forms attached, some 20 details of ingots cast were to be recorded, with some aspects associated with bonus payments to the men, for example, satisfactory filling and plugging. It was a lively experience for they would always claim success in achieving what was required, against which it was difficult for a 20 year old to stand. 'Flying stoppers' were also a problem. If the metal stream could not be shut off by operation of the lever which lowered the stopper into the nozzle, the ladle had to be moved on to the next mould with the steel still leaving the bottom of the ladle and having to traverse the two horizontal mould edges and the ground between. The impact of the stream on the mould edge produced a horizontal spray of coarse molten steel droplets at about ankle height which had to be avoided. In addition, the 'young lad' was subjected to torment by the crane driver in his cabin above, who would wait until the shift was nearly over, bring his crane with open gearbox just overhead and jerk the trolley back and forth, spilling black oil over person and clipboard. That meant that another hour or two had to be spent making a fair copy, hopefully accurate, of the records for submission to the melting shop office. One just had to grin and bear it without other than minimum comment until the novelty wore off.

All in all, the Corby experience was an extremely valuable one. I learnt a very great deal and in later years when working in the steelworks environment for the British Oxygen Company felt very much at home. It may even have helped me teach ferrous extraction metallurgy in a more interesting way years later.

Third Year 1946–47

This was a most enjoyable period. The staff in the Metallurgy Department at the RSM were, on the whole, very good at their job and helpful, providing an interesting course which gave a good, all-round metallurgical education. Where the subject had moved forward rapidly beyond their immediate interests and expertise, outside lecturers were brought in. Two examples: Hume-Rothery, from Oxford, gave a series of lectures on the then new electron theory of metals. In an already difficult field we found his lectures hard to follow, particularly since his total deafness made voice control, in terms of volume and pitch, uncertain. H.J.T. Ellingham, who had published the first papers with Professor Dannatt on the free energy concept of metallurgical reactions, covered this in lectures to us. We were probably the first group of students to be taught metallurgical thermo-chemistry with the aid of 'Ellingham diagrams'.

Dr Ellingham was essentially an electrochemist. Born in London, he gradu-ated at the Royal College of Science in 1917 with a First class honours degree in chemistry and, after a short spell of military service in Mesopotamia, in 1919 he returned to the RCS as a Demonstrator, eventually becoming Reader in Physical Chemistry in 1937. A task he undertook was to teach the elements of physical chemistry to chemical engineers and to metallurgists in the City and Guilds College, and the Royal School of Mines respectively. This was to bring him into contact with Professor Dannatt at the RSM, whom I firmly believe was instru-mental in encouraging Ellingham towards a systematic consideration of sulphide and oxide formation in the context of roasting and smelting reactions. Dannatt may well have been responsible for inviting Ellingham to give the lectures and, in any event, there can be no doubt that their association, exemplified by their joint paper to the Faraday Society in 1948 (Discussion No. 4) 'Roasting and Red-uction Processes – a General Survey,' was the reason that Ellingham became particularly involved with metallurgical interests, culminating in his chart pres-entations of free energy change for reactions under standard conditions over a range of temperaturs, which are contained in the paper. His graphical presenta-tions greatly assisted understanding and in this respect can be likened to those developed by Professor Michael Ashby at Cambridge for the properties of engi-neering materials many years later.

Ellingham's lectures were excellent, although since he had spent the previous hour drawing all the graphs and equations he was going to use on the blackboards, we were hard pressed to both listen, note and copy the material on the board. The use of visual aids, with copies handed out, would have saved him time and improved our appreciation of the lecture contents, which have since become the standard approach to metallurgical thermochemistry and have greatly facilitated the teaching of subsequent generations of students throughout the world.

A great deal of effort on metallography continued, aided now by the presence of Dai Lloyd Thomas who had left Stones and joined the RSM staff. The physi-cal metallurgy of steels was taught, again effectively, by Mr Ron Harris where a mark of the timely relevance of the course was the considerable attention given to hardenability theory, in particular to the work of Grossman. Later in the year students were requested to give a lecture on a topic chosen from a provided list and my choice concerned hardenability. It was extremely fortunate, therefore, when an essay question on this topic appeared in one of the final examination papers! Non-ferrous alloys and general physical metallurgy, dealing with phase diagrams and industrial alloys fell to Dr Munro Steel Fisher. He was a short, quiet Scot, known to his colleagues as 'Jock', who taught extremely well, with carefully prepared hand-outs. Mechanical treatment, mechanical testing, materials properties and material selection was taught, somewhat ineffectively for me, by gentle Monty Chapple, who in the latter aspect simply wrote up British Standard Specifications on the blackboard. His major attempt to make mechanical treatment interesting was to describe in some detail how the first reversing rolling mill was driven by a steam locomotive with the wheels removed, some 100 years earlier, all related in monotone with eyes fixed on the ceiling! Chapple also lectured in electro-metallurgy with perhaps a little more success. As in all university departments the atmosphere was much affected by the attitude of the technical staff. At the RSM there was an excellent relationship, particularly with Bill Arkill and Bill Chivers, the latter in charge of the workshop.

During the 1947 Easter Vacation, there were two major expeditions. The whole class went by coach for a metallurgical tour of Sheffield (visiting Samuel Fox - Stocksbridge, Firth Brown's Atlas Works, Park Gate Iron and Steel Co. Ltd., Sheffield Smelting Co. Ltd., Firth Vickers Stainless Steel Ltd., Hardy-Pick Ltd., and Hadfields East Heclar Works) and the Liverpool area (visiting Thomas Bolton - Widnes, High Speed Steel Alloys Ltd., John Summers Ltd., White Cross Cable Co., - Warrington, British Insulated Callendars Cables Ltd., Rylands - Warrington, British Aluminium – Warrington. Sadly most of these companies no longer exist). Few departments today seem to operate such substantial tours, although I am delighted to say that the tradition continues at the RSM. It is even difficult now to place students into worthwhile vacation jobs.

In 1947, conscription of young men for the armed forces was still in existence and at about Easter time we were interviewed by representatives from the War Office to assess our potential, with advice given to them by the staff as to our academic achievement. It seemed that I was destined for an Upper Second class degree and I was informed that on that basis, I would be required to join the Army. The year 1946–47 was also marked by the appearance in the Department of my old school friend, David West, who had graduated with a First class degree from Cardiff and commenced research at the RSM on the solidification of horizontal steel castings with British Iron and Steel Research Association support, the castings being made by the English Steel Corp., Ltd. His work was to be relevant to some aspects of my own many years later, on the structure of ingots. He also established in my mind the value of Oberhoffer's reagent in revealing phosphorus segregation, of great importance in the study of early iron artefacts in my later archaeometallurgical work.

Finals, June 1947

In spite of the knowledge that I was to be called up on graduation, preparation and revision for the final examination was vigorous. In those days, considerable emphasis was laid on practical examinations, with four-day assessments in both metallurgical analysis and in metallography. In the former, I recall having to determine the magnesium content of dolomite, amongst other questions. This required double precipitation as magnesium pyrophosphate with standing over-night in between. Thus on the third day it was crucial not to drop the beakers! Sadly, in my view, practical examinations are now a thing of the past, the main excuse being that they are unpopular with the students!

The external examiner was Professor Higgins from Swansea University. The corporate view was that whilst he was an expert on steels, he was not particularly *au fait* with the metallography of aluminium alloys and preferred to talk about the former. In the Metallography Practical several alloy specimens, both ferrous and non-ferrous had to be polished, examined, described and specific questions answered regarding their condition. In addition instructions had to be given to a welder for the formation of a joint between two given steels and the result as-sessed metallographically. Professor Higgins would wander round assessing progress and asking questions. On each occasion he approached me, I arranged for an aluminium specimen to be the subject of my scrutiny, but on the third day in the afternoon he perched himself on the bench beside me saying, with suit-able robust emphasis, "I know that you know that I don't know much about aluminium alloys - put steel specimen A under the microscope!" In a later per-sonal interview he quizzed me on one particular script where I had "gone into cuckoo land" concerning malleable cast iron, something I had recognised as soon as I emerged from the examination hall, to my disgust.

The upshot of all this was that, contrary to the expectation declared to the War Office at Easter by Professor Dannatt, I was awarded a First Class rather than the Upper Second expected. This created something of a problem. It had been agreed that Firsts would be reserved for research/industry, while the others would be called up. In my case anyway the Army had been particularly keen to have me because of my War Certificate B and earlier training, which would have given a commission directly before the war. I was, in fact the only member of the class still active in the Senior Training Corps, as a volunteer. They moved smartly to effect my entry, with a call-up and orders to attend the Officer Selection Board at Haslemere shortly after the end of term. This lasted for four days in which both written and diagrammatic field intelligence tests and leadership tests were un-dertaken, with several assault course exercises to make sure we were fit! At the end of one such exercise on the fourth day, trembling with exhaustion, I was called to the Regimental Sergeant Major in charge who said "I have this telegram for you *Sir,* I have kept it until the end since I didn't want to spoil your fun!" It simply stated the I had been released from the Army to essential work under an order from the Ministry of Labour. The Department had been busy with J. Stone and Co., who, of course, knew of me and needed a metallurgist in the Research Department and had submitted the request for my release. Thus ended a very short military career.

In an interview after the aforementioned assault course I was told that I had passed with good marks and would have been accepted for a commission and that the Army was sorry to lose me. I recall being completely bemused by the whole affair, not knowing whether to be glad to not, but in the event not protesting at the course matters had taken. As already indicated, most of my year were called up, with several ending up at the Imperial Smelting Corporation afterwards. In May 1948, I attended what I believe may have been the first ceremony at the Albert Hall for the presentation of degrees after the war, with hundreds of others from the many constituent institutions of the University of London.

Industrial Employment

J. STONE AND CO., LTD., 1947–50

I had already worked at Stones as a vacation student, as previously mentioned, so I was very much 'at home' in the Research Department at the Deptford works from the start. Work commenced at 8:30 a.m. and ended, if my memory serves me, at 5:30 p.m. On Saturdays attendance was required in the mornings and there was always a rush to get away at lunch time for whatever weekend activity had been planned. By this time A.J. Murphy had been made a director of the company and R.J.M. Payne had taken his previous role as Research Manager.

Rosenhain's connection with the company as a consultant after leaving the NPL in 1931 was evident in that one of the microscopes was still known as Rosenhain's. Walter Rosenhain (1875–1934), who had been an 1851 Exhibition scholar at St John's College Cambridge in 1897, had established links between J. Stone and the NPL, where he was Superintendent of the Metallurgy and Metallurgical Chemistry Laboratory from 1906–31, during which time, amongst numerous alloy constitution and metallography achievements, he designed a metallurgical microscope. On retirement from the NPL in 1931, he became an independent consultant, mainly to J. Stone, spending much of his time in the laboratories there. There seems little doubt that the microscope in question, which only he was allowed to use during his consulting years, was of his design. I wonder what happened to it eventually? The link thus established by Rosenhain was to result in several NPL staff moving to Stones in support of the coming war effort. These included A.J. Murphy himself, S.A.E. Wells and R.J.M. Payne.

The Research Department was entered through a formal hallway, with the research foundry (Fig. 7) and workshops to the left on the lower floor and offices, metallography and general research laboratories above. I was given a desk in one of the glass-partitioned offices, sharing with A.W.O. Webb ('Alf') and another Welshman, Bernard Morgan. With two Welshmen in the office, I always had a hard time when England failed to win at rugby! A teenage assistant, Phil Webster, later emerged as a lecturer at the National Foundry College. He has made valuable contributions in the field of methodising castings through computer simulation, and has recently had projects at Birmingham University IRC, Casting Centre. In relation to mould preparation in the research foundry, 'Stocky' Stockdale, an elderly moulder from the North-East, taught us much from the wealth of his experience. The practical skills that he demonstrated were of great value to me later.

*Fig. 7 Gas fired pit furnaces in the Research Department Foundry, J. Stone and Co. Ltd.,
Deptford, 1948.*

Alf Webb's particular area of activity was propeller bronzes, being associated
with the development of the 'Superston' copper–manganese–aluminium alloys
(CMA) from the high tensile brasses. He was subsequently to hold important
positions for Stones as Chief Metallurgist and then a Director, with a spell as
Managing Director of Stones in Brazil. After retirement he took up new work
with Lloyds Register of Shipping. Joe Eynon, Les Gwyther (more Welshmen!)
and Payne had responsibility for the development of magnesium alloys, with
emphasis on creep-resistant grades. My only effort in relation to magnesium was
to report on the condition of castings for CARD (the bouncing bomb developed
for use on land, to enter hangers etc.) after dropping from aircraft with impact
onto concrete. By a strange coincidence, a colleague in the Cambridge Department during the later years, I.M. (Ian) Hutchings, wrote about the 'bouncing
bomb' and its use. Ian, also a Fellow of St John's, initially had no idea of my
involvement.

Bearing Metal Research

My main responsibility was for the development of tin-base bearing metals
('Babbitts') and for control of the foundry office and stores. This was made difficult by the technicians' lunch-time activity in the research foundry of a 'homer'
production line for cast 'lead' soldiers, using two-part metal moulds. The output
was hand-painted, mounted in specially-purchased boxes and sold on Petticoat
Lane. The Babbitts were ideal alloys for the purpose, as they were stronger and

harder than lead. There was great difficulty initially in preventing the scrounging from the store of the valuable tin stock as well as of the ingot supplies of various proprietary bearing metals. Not in authority over those involved, the first thing to be done was to secure the valuable tin and antimony stock by running it down and providing a lockable cupboard for the minimum needed. Quite quickly the 'producers' were obliged to bring in their own lead scrap (presumably from redundant plumbing!), which was not as good for the purpose, and gradually the novelty wore off and the problem was resolved. Interestingly, there was never any attempt to filch test bars or specimens from my research work.

The main thrust of the work on tin-base bearing metals was to improve the fatigue performance by a reduction in grain size. Although making serious attempts to predict the degree of solubility of other elements in tin and the likelihood of compound formation (where small additions of another element might produce a nucleating compound) using Hume-Rothery's rules, it has to be admitted that the approach eventually became one of adding much of the Periodic table! A standard series of castings were produced for each alloy, representing different cooling rates: sand mould, iron and aluminium moulds and wedge chill. Whilst silver alone gave some grain refinement, the most significant results for the tin–antimony basis alloy came from the addition of small amounts of cobalt and silver together, where the normal columnar structure was changed to very fine equiaxed, even in the slow-cooled casting. The effect increased with antimony content in the range 5–10%Sb, with the greatest being at 10%Sb and no copper (Fig. 8). It seems likely that a peritectic cobalt–tin compound was nucleating solidification with the eutectic-forming silver reducing growth, although the refinement mechanism was never determined. Throughout the work it was, of course, necessary to cut and polish specimens for micro- and macro-examination, several hundred of them, and the soft matrix bearing alloys are difficult to prepare!

Another route to grain refinement was by the application of vibration to melts during solidification. Ultrasonic vibration using a magnetostriction oscillator, with a probe mounted on the end of the nickel oscillator tube dipping into the melt, gave a very fine structure, but limited to the zone around the probe. At the other end of the scale, mounting moulds on oscillating tables, putting in large amounts of power at low frequency, resulted in metal ejection. Effective modest refinement was achieved by a compromise. By both routes, mechanical testing indicated that some advantage in properties had accrued, but there seemed to be no enthusiasm on the part of management to take the matter further. In fact, soon after I left, in 1950, the bearing metal side of the business was discontinued.

During this work on grain refinement we were concerned with the influence of thermal fatigue in tin-base bearing alloys, where the anisotropic expansion of the tetragonal material produced cracking at grain boundaries and at interfaces with second-phase compound particles such as Cu_5Sn_6. Through this connection,

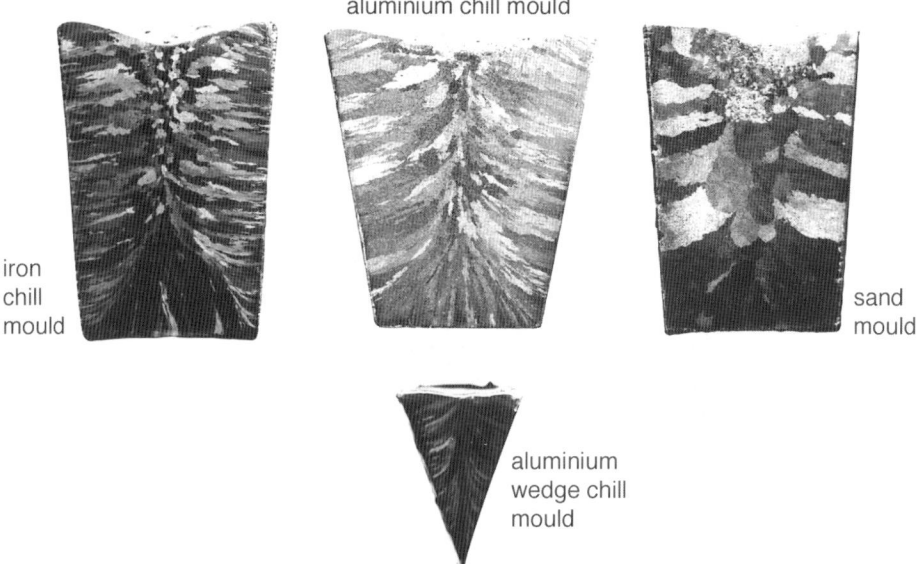

Fig. 8a Macrostructure of standard 86Sn–8.5Sb–5Cu bearing alloy castings
(composition in wt-%).

I came across the work of W. Boas and R.W.K. Honeycombe in Melbourne, Australia, the latter to become my Head of Department at Cambridge in later years. There was also to be a link to much later research at Cambridge on the thermal fatigue of solder bumps in flip-chip joints, described later.

Not all of my effort at Stones concerning bearing metals was on the laboratory scale. I recall lining the stern tube of a directors' yacht and developing a method for building up pads on a very large shell using cast alloy sticks and a welding-on technique. One piece of work, which was taken to full fruition, concerned the preparation of spectrographic standards for bearing metals. Control analysis was by emission spectroscopy, comparing line intensities for production with those obtained from standard specimens, bracketing the permitted range of impurities. Several of the impurities, particularly iron and zinc, segregated badly where 'standards' were prepared by casting, with very different values obtainable from end to end of the rods used, however rapidly these were solidified. The problem was overcome by spraying the alloys from above the liquidus into powder and then compacting the powder into short cylinders. The gross strength of these, without any binders, was sufficient for machining and handling in the analysis, and although there may have been segregation on a very fine scale within each powder particle, all particles were of the same overall composition and were very small compared with the sparked area. Thus, very consistent values were obtained throughout the life of the standard bars as they were used, with surfaces renewed

85.3Sn–5.0Sb–6.5Cu–3.0Ag–0.2Co

83.3Sn–8.5Sb–5Cu–3.0Ag–0.2Co

86.8Sn–10Sb–3.0Ag–0.2Co

Fig. 8b Castings showing grain refinement of tin-base bearing metal by the addition of silver and cobalt and influence of antimony content. (scale ×0.64)

between sparkings. This technique was later also adopted for bronze by B. Doig, after he left Stones for the British Non-Ferrous Metals Research Association.

A particular penalty of this period for me which is, I suspect, common for juniors in industry, was that there was never any suggestion that any of my work should be published. Thus there is no record, apart from this belated and brief account, for subsequent workers. Small wonder that junior academic posts, where publication is a strong recommendation, are most open to postgraduate students where a three-year PhD can lead to several papers. Thus, staff movement between academia and industry, usually considered desirable, is made more difficult. This is not a criticism of Stones. It may well have been judged that the work was not at a suitable stage, and certainly some of the work of others, more senior, was published, particularly in relation to magnesium alloys and propeller alloys.

R.J.M. Payne was not any easy man to work for. He was then rather acerbic and dour but after a late and very happy marriage, he was greatly changed. I only met him once subsequently and was struck by the transformation. A particularly infuriating feature was his insistence that he alone had the required command of the English language. The first draft of an internal report would be torn to shreds, a process repeated on the second draft when his first amendments had been made, regenerating much of the first format! It is a matter of interest that Payne's secretary was male, a Mr Fisher, who also taught shorthand and typing at a local evening class. Prior to the Second World War, and certainly in the 1920s, male secretaries were more usual than women, and Fisher was a carry-over of that situation.

J. Stone and Co., Ltd., had always had a strong research team. During the Second World War, they had faced many problems where the work on an order could be set in motion or brought to a stop by the availability or non-availability of materials. In the 1940 Blitz, important sources of supply could disappear overnight. As an example, a Bomber Command demand for a big increase in incendiary bomb production (Stones produced 28,300,000 of these during the war) could mean that a useful saving in the weight of an aircraft component by the substitution of a magnesium for an aluminium alloy could not proceed in the face of the overall supply position. The metallurgist's task, amongst others, was to select suitable alternatives for whatever materials were in short supply, without any time for prolonged testing of the substitution. The selection of a hardened aluminium bronze for ball races in place of chromium bearing steel was a case in point. In this setting, and with the encouragement of Government agencies, there was always mutual assistance between companies, not only in matters of supply but in sharing technical information. J. Stone under Murphy had a strong metallurgy research base and were a leading force on Government committees controlling the metallurgical war effort. They were active in the development of new alloys, for example the Superston range of marine propeller alloys, and eminent in the production of aircraft high quality castings in both magnesium and aluminium alloys. However, much of their research time during the war had been taken up with finding solutions to other manufacturers problems, all associated with advanced military applications of alloys, giving far more help than they received. With the end of the war and the sharp fall-off in military orders, such domestic outlets for aluminium alloy castings as presented themselves usually did not require the degree of metallurgical control that had previously been in place, and some retrenchment and changes of research policy were clearly called for. This resulted in a concentration of magnesium alloy and propeller alloy research in a new especially built laboratory at Charlton not long after I left. All that now remains of the Stones site there is a light metal foundry. All propeller castings are now sourced outside the UK and assembled here.

By 1950, my Ministry of Labour 'Bevin-boy type conscription' was over and I was free to move and needed to raise my salary. It had changed from, I think, £365 to £375 per annum in three years. Living costs were rising and there seemed to me to be limited opportunity for advancement at Stones, with several young men ahead of me in the organisation. I therefore answered one or two advertisements. One was for a post at ICI Blackley, Manchester, now the home of Zeneca, with the remit of plant maintenance. One of the interviewers had known my father from his ICI Ardeer days. I was offered the job, but realised that a great deal was to be concerned with corrosion problems, and, unlike Cambridge graduates of those days, I had had virtually no training and certainly no experience in the field. In recent years I have been back to Zeneca as a consultant, where one of my research students, Laura Cohen, now holds an important post in the Engineering Department dealing with, amongst other things, plant corrosion problems! The post I eventually took was at the British Oxygen Research and Development Company (BORAD), in South Wimbledon, at £540 per annum.

THE BRITISH OXYGEN COMPANY, 1950–60

The research centre for BOC was on an extensive site in the Lombard Road factory estate, reached by a slip road from the main road between South Wimbledon Station and Morden, and over the years of my time there grew appreciably, taking in the buildings of other smaller companies as they became vacant. It was not a straightforward journey from my home in Bromley, Kent: bus, train, underground, bus/walk, taking about $1^{1}/_{4}$ hour each way.

In the early days, the Gas Applications Group, which I had joined, was housed in prefabricated single storey buildings, rather better than 'huts' but still very utilitarian. The group was managed by Dr L.C. Bannister, a short man with jutting chin and the son of metallurgist Professor Bannister of Liverpool University.

Bannister had obtained his PhD at Cambridge under U.R. Evans before World War II and had then worked for British Insulated Callender Cables Ltd. in the North West for a number of years before joining BOC. All those in charge of programmes were required to meet him formally in his office weekly to report. All the section leaders dreaded these occasions, not that he was in any way unpleasant, but he could be doggedly illogical and it was difficult to maintain a calm and polite attitude, as required by the very formal nature of the meetings and by his generally kindly disposition.

Oxygen Cutting

On arrival, I was attached to Dr Robert Kerr who was in charge of oxygen cutting research. I shared an office with Dr George Roberts, a physicist who had worked for his PhD with Professor Ubbelohde at Queens University, Belfast, on the effect of hydrogen on internal friction in palladium. He was an extremely bright young man, but sadly died of cancer not long after.

George was involved with considerations of gas pressure regulators and gas jets from nozzles in relation to the design of equipment, a general area in which other graduates, Eileen Sheriff, John Hardy, Fred Williams and Denis Milner were also involved. Den Milner moved on to an academic post in the Metallurgy Department at Birmingham University, eventually becoming a Reader. John and Fred continued with BOC, moving into commercial activities once research had been cut back.

Bob Kerr was not an experimentalist and his approach to oxygen cutting research was confined to measuring gas flows through torches with different types of nozzle and different oxygen/acetylene mixing systems. There seemed to be a timidity in relation to actually lighting torches, let alone using them in their function of cutting steel. The experimental cutting rig consisted of a heavy, moving table on which a steel section could be mounted, set beneath the selected cutter fixed to an overhead beam. The drive for the table was by lead screw powered through a Carter infinitely variable gearbox. This arrangement gave very precise and steady motion, so important to investigation of the process. The fixed position of the cutting head facilitated measurements etc. in the reaction zone. This equipment had never been used and, as may be imagined, for me this approach was very dull and restrictive.

Although oxygen lancing and the derived process of oxygen-cutting was long-established, very little fundamental investigation had been carried out. Dr Alan A. Wells at The Welding Institute, Abington had studied the combustion of rotating rods of different metals and alloys in an oxygen jet and some limited work had been carried out by Slottman and Roper in the USA. In the process, steel is raised to the ignition temperature by an oxyacetylene flame and then subjected to a jet of oxygen. Since the melting point of the iron oxide, at about 1350°C, is less than that of steel, it can be swept off the reaction surface, enabling surface removal as in scarfing, or a slot can be produced with a jet at 90° to the metal surface in cutting.

Sadly for him, but as it turned out helpful for me, Bob Kerr injured his back at home and was off sick for a period of about three months, during which time I took over the section and started cutting in earnest (Fig. 9). The impact on the assistants of the flying sparks and slag was dramatic and enlivening! On his return, Kerr was moved to other duties in the Gas Applications Group and finally became Patent Officer for the whole research centre, which suited his preference for desk work much more closely. I took over the cutting research for a total of three years. Considerable fundamental insight was gained into the cutting process itself, the effect of oxygen purity, the influence of nozzle design etc. It became clear that, chemically, the process was extremely efficient and that, for example, the increased cutting speed achievable with a convergent/divergent (venturi) nozzle was the result of the oxygen jet being narrower for a given flow rate of oxygen. Thus, there was less steel removed in the cut (kerf) per length run and in turn, for a given oxygen flow and time, a greater length could be severed. Sensitivity to oxygen purity, and to the density of the impurities present in the cutting oxygen, clearly indicated a gaseous diffusion process existing over the boundary layer at the oxygen/metal interface, with cutting speed mathematically relatable to the level of impurity and its diffusivity. The work was most rewarding and remained definitive even 25 years later when BOC invited me from Cambridge to talk about it at the Faraday Conference on heterogeneous oxidation in Leeds, leading to publication of a review in *Metals Technology* (1978, **5**, (5), 163–175). Very recently data from reports on the work carried out in 1950–53 have been used by BOC in a modelling investigation of the cutting process, in cooperation with the RSM, as a PhD project.

During my time as leader of the Cutting Section of Gas Applications Research a team developed with an ex-RSM metallurgist, C.A.F.T. Spray (Alan), as my number two and three research assistants, Keith Latimer ('Chiefy', ex Navy), Bill Eley and an Australian, Ian Webb. They were a lively bunch and work was good fun. An impression of this can be gained from the following verses dedicated to me when two of the assistants left the group.

'Parting Shot'

Are you filled with disgust when your spark gun you've bust?
And your nozzle's been chewed up and bent?
When your cut is all drag and your goggles all slag,
And your turn-up is charred to a vent?

Fig. 9 Oxygen cutting at BORAD; JAC with assistants Keith Latimer and Bill Eley.

Are you less than amused when a pun's being used
And you're asked for some ninety-four kerfs?
When the spray-gun is screaming (with DMS beaming)
And your RA's are looking like serfs?

Though the drone of his arc-weld may sound like bad jazz
Though the freeze-grind and future look bleak,
No one need worry, not even 'Old Chas',
We two departed last week.

Chiefy, Bill
With apologies to *Punch*

Some words of explanation may help! The spark gun was required to light the cutting torches. 'Drag' was the extent of departure of the line of the cut from the vertical; with more than a certain amount of drag, drop severance would not be obtained. 'Ninety four kerfs' would have been cuts (slots) made with a No. 94 nozzle whose width could then be measured. Metal spraying and the development of a freeze-grinding process for bones reflect the variety of investigations taking place in the area. Alan Spray also left soon after, to train as a Church of England minister. He is now Canon Spray of St Andrews Church, Burgess Hill.

By 1954 it was apparent, however, that oxygen cutting research had to stop. The limitations of further development were now clear through understanding of the process, whether this be in the cutting of thin sheet or the severing of heads from large forgings several feet thick or in the surface oxygen gouging operations such as scarfing and deseaming, so closely allied in nature to the cutting process. New premises had by then become available with extension of the BOC site into an adjoining vacated property. This provided Gas Applications Research with a road-fronting office block, metallography laboratory, welding laboratory and a major area committed to investigations in the use of industrial gases in metallurgical processing, particularly iron and steel making, which was to become my primary responsibility. Some of my colleagues at the time may be seen in Fig. 10. Although not in this (unofficial) photograph, by 1955 the Gas Applications Research was managed by a Christ's College, Cambridge, graduate (1940) W.J.B. Chater (Fig. 11), in succession to L.C. Bannister, the latter having been promoted to General Research Manager. Chater had previously worked in the Gas Separation Group at Grafton Street, then headed by Dr P. Schuftan.

Bill Chater was a remarkable and talented man. During the last war, not long after graduation, he was a member of the team that sabotaged Germany's heavy water plant in Norway, having been concerned with hydrogen production at ICI for a few years, and presumably providing the expertise to identify the critical components of the plant. He was a good natural scientist, able to bring soundly-based argument to bear and always encouraging and supporting those that worked for him if they gave of their best. Although, in stature rather large, he had a kindly, gentle disposition and was admired by us all. A particular memory is of a returned draft report which had his approval apart from axes labels missing from a graph on which he had scrawled "millifirkins per lunar year?" On one occasion he asked what I was taking home in my briefcase one Friday evening. On being told that it was a report that I wished to finish he commented "either I am giving you too much to do or you are not up to the job." He placed proper emphasis on family time and on Parkinson's Law.

Sadly, however, BOC higher management did not seem to rate him. When Gas Applications Research and much other research at the Morden site was closed down some years after I left, he was required to act with Sales Technical Service, based in Leeds. His home was in Surrey, but he existed in a poor rented room in Leeds for some time. He then became Training Officer at the London headquarters, Great West House. I attended his eventual retirement party there some years later but felt that there was no real recognition of his contribution to the company as a scientist. Those presiding only seemed to know of his final years in the organisation of training. His retirement was short lived before a fatal heart attack, which saddened me greatly.

Fig. 10 Gas Applications Research Group, BORAD, showing Arthur Bracken, Dick Jahn, Fred Williams, Laurie Brittaine, Norman Crevis, Pad Arti, Ken Sargent, Bob Cresswell, John Hardy, JAC, Maud Tabor and May Williams with others.

Oxygen in Process Metallurgy

Once committed to research into the use of gases in metallurgical processing, particularly oxygen in steelmaking, liaison naturally developed with J.L. Harrison (Jack), who was at that time Technical Adviser, Iron and Steel Processes, British Oxygen Gases, based in Leeds. Born in Epworth, Lincolnshire in 1910, he had been technical assistant to the works manager at the steel foundry of John Brown and Co., in Scunthorpe, becoming steel plant manager responsible for cupola/converter and electric steelmaking facilities. After the war, in 1946, he moved to a similar position at Catton and Co., Leeds where he successfully developed the regular use of oxygen in side-blown converters (Tropenas), which could well have been the first real application of oxygen in steelmaking anywhere, although Harry Everard of Edgar Allen and Co. maintained that he also was 'the first to use oxygen in Tropenas converters'. In any event, both may have been stimulated in their efforts through their contact with the British Iron and Steel Research Association, since in December 1945 D.J.O. Brandt (Oscar; later to be Editor of *Iron and Coal Trades Review* and something of a thorn in the BISRA establishment before an untimely death) witnessed the conversion of 30 cwt of cupola metal by an oxygen jet projecting onto the surface, melted and held in a $3^1/_2$ ton electric arc furnace, I think at Brymbo Steelworks. The sad thing is that, as with so much innovative work in the UK, proper development of it to larger

Fig. 11 W.J.B Chater.

practice took place elsewhere. With normal air blowing in side blown converters there was a requirement for high silicon irons to generate adequate heat, and the use of oxygen gave much greater thermal flexibility and a wider choice of charge, particularly important in units associated with cupolas for which cheaper steel scrap rather than pig iron could be a major charge constituent.

In 1949, Jack moved to Low Moor Alloy Steelworks (Bradford) as Technical Adviser to the Managing Director, continuing with the development of oxygen usage for assisted melting and decarburisation in electric arc furnaces, and contributing substantially to work that led to the first continuous casting of high speed tool steels. In 1951, he joined BOC in the post described above. His role was to explain, to the steelmakers particularly, how the application of oxygen could bring them benefit and to persuade the process operators to agree to trials to prove the point. He was adept at presenting technical arguments to win over customer involvement. At that time in the 1950s, BOC's endeavours were largely in relation to the improvement of performance from existing processes, e.g. flame enrichment and lancing for decarburisation in open hearths and decarburisation in electric arc furnaces. The collaborative trials usually required sustained attendance by BOC representatives (and 21 shift operations in steelworks was then the norm), as C.N. Walters, Jack's personal assistant recalled somewhat wearily.

It would, perhaps, be tedious to call up detailed memories of these many trial events in steelworks, but two in particular remain very clear. The BOC main board Director responsible for research, had met a Director of Steel, Peech and Tozer ((SPT) later part of United Steels) and had apparently convinced the latter that oxygen flame enrichment would be likely to be effective in increasing output from the Rotherham open hearth melting shop. I was instructed to co-ordinate trials, with 24 hour observer coverage over a month on one basic open hearth

Fig. 12 Converter ladle for top-jetting experiments.

One major piece of research carried out in the BORAD days concerned the use of argon through hollow electrodes in electric arc furnaces, using the Birlec single phase twin electrode, 1 cwt direct arc furnace, where introduction of argon in the arc reduced arc resistance, enabling melting to be achieved more quickly with reduced noise, fewer short circuits or open circuit breaks and reduced electrode consumption. Melting was carried out on an almost daily basis by Charlie and Tony with a foundry worker, and they took the main load of the works trials of the idea, again at SPT, Rotherham, but this time in the electric melting shop where Richard Howes was melting shop manager. Dick's uncle, Clarence Howes, had been my father's best man in 1912, when both were young men in Cambridge. Clarence and Dick's father, Sidney Howes, were the sons of John Howes, manufacturer of the Granta bicycle at their premises in Regent Street. Whilst Clarence stayed on in Cambridge, eventually running the cycle shop, Sidney moved to Sheffield, where he eventually became Managing Director of Samuel

Fox, Stocksbridge. He and his son Dick Howes lived next door to each other in Dore and Totley and were visited regularly by me when engaged in trials in Sheffield.

The use of argon through hollow electrodes at SPT certainly resulted in improved operation and electrode consumption was reduced, but compared with the hand-operated Birlec laboratory furnace at BORAD, the automatic electrode control mechanisms reduced the significance of the reduced arc resistance as related to arc length and the lower frequency of arc interruption. However, from time to time the idea re-emerges in literature and may have found some application. The Birlec installation set up for hollow-electrode work, with a powder dispenser and a line to an electrode, for making additions, is shown in Fig. 13.

Another interesting idea which went to plant trials was the use of electroslag hot-topping where the surface electrode was a Fusarc type wire fed and melted into the top of an ingot. By control of the overall wire composition, to include the additives encapsulated by the multi-stranded wire, segregation in the head of the ingot could be compensated.

A number of overseas visits took place during this period, including a European tour with Jack Harrison, taking in Donawitz at Leoben in Austria, where the LD oxygen-steelmaking process had originated, a conference at Reutte in Tyrol, the home of Metallwerk Plansee, producers of hard metals, carbides etc. and the furthest trip, to India in 1959. I was required to go to assist the Indian Oxygen Company in the promotion of oxygen usage in the newly emerging and traditional steelmaking centres. Visits were made to the National Metallurgical Laboratory and their low-shaft furnace project, which was much in sympathy with world-wide interest in the use of low-grade raw materials, but in my view were then unlikely to be economic. Tata Engineering and Locomotive Co. Ltd., the Central Fuel Research Institute and the Kulti and Burnpur works of the Indian Iron and Steel Co., Ltd. were also visited, with advice given on general steelmaking problems and possibilities for oxygen applications, particularly in raising metal temperature to avoid skull formation in transfer ladles and the potential for a change from duplex converter/open hearth to LD practice. The National Metallurgical Laboratory (Director, Dr B.R. Nijhawan) had also organised a conference at Tata Steel, Jamshedpur, at which I gave a paper on desiliconisation, which was still an important topic in India, in a session chaired by Dr T.P. Colclough. In the paper, the opportunity was taken to introduce our work on a continuous concurrent desiliconisation process in which iron poured down a tower reacted with injected oxygen, taking advantage of the dispersed state of the iron. The $1-1\frac{1}{2}$ cwt scale of our work is indicated by Fig. 14. Such a tower could be inserted as part of a pouring system. The representatives of Indian Oxygen who had looked after me so well during the visit were A.K. Gupta, A.K. Bettacharya, S.M. Ray, R. Bhandari, and P.K. Dam.

Fig. 13 The 1 cwt electric furnace with Charles King at the controls.

The staff of the research foundry changed somewhat during the 6–7 years before I left. Cedric Walters (now deceased) was with me for a while initially. He was a graduate from Cambridge and as a student had invited me to speak to the Goldsmiths Society there after a spell with me as a vacation student. He was quite quickly moved to Leeds to assist Jack Harrison as mentioned earlier. My next assistant was A.G. Cowen (Tony). He had graduated from the RSM and stayed with me until I left in 1960. Later, he was to work with me at Cambridge as an Industrial Fellow, for his PhD. T.C. Churcher (Tom) also joined in my last year or two at BOC, and took over some of my projects when I left. Tony Cowen eventually became Research Director of Telcon Metals before retirement. Other notable workers associated with process metallurgy at BORAD were Leslie Cook, John Harding and Kenneth Sargent. The latter made important contributions to the understanding of fume formation and its control.

Fig. 14 A desiliconisation tower.

An early requirement of my role at BOC was that I should provide a series of lectures on extraction metallurgy, welding and mechanical treatment for all the Sales Technical Service staff and engineers involved in the expanding steelworks customer base. These were held in the BOC plant engineering office in Grafton Street, London, W1. One of those attending was an ex-steelmaker recruited from Appleby–Frodingham, A.N. Whiting (Arthur). He was to go to South Africa quite quickly in a Sales Technical Service capacity to develop oxygen use in extractive metallurgy there. In later years, when he was Managing Director of Davy United, he and I were to become good friends when we were President and Vice-President of the Metals Society respectively. During 1980–83, we represented the Society in the negotiations to amalgamate with the Institution of Metallurgists to form the new Institute of Metals. At much the same time as these British Oxygen Company lectures I became involved in taking occasional evening classes at the Sir John Cass Institute (part of the City Poly) and giving a regular Higher National Certificate (Mechanical Engineering) evening course at the Wimbledon Technical College with both lectures and practicals on metallurgy for engineers. Looking back, the syllabus was impressive and quite demanding. I can recall mounting my first practical demonstration on the subject of corrosion, with two beakers containing salt solution, a salt bridge between them and a steel plate in each connected to each other through a needle galvanometer. With oxygen bubbled through one beaker, the current generated passes one way and in the reverse direction if the bubbling is switched to the other. Thus, the evidence of differential aeration, establishing anodes where air/oxygen cannot penetrate (as under damp mud and seams in cars), was provided to the obvious interest and almost disbelief of the students present. This teaching experience was invaluable in relation to the decision that had to be made later in my career.

There was a generally happy atmosphere in the Gas Applications Group of BORAD Ltd. under the management of Bill Chater. There was even a small choral group that gathered in the lunch hour, using a hired room in Wimbledon, under the leadership of the aforementioned Ken Sargent, a chemistry graduate from Imperial College. Having learnt of our activities, the bosses of BORAD sometimes requested a performance from the group at the annual dinner. Another member of the group was the welding metallurgist Bob Cresswell and at one such dinner he and I sang 'Let the Grass Grow' from the Julian Slade musical *Free as Air*, as a duet. The lyrics concern the 'busyness' of men and commend a more leisurely approach to life. Somewhat lacking in a sense of humour, one of the top table complained at our choice of song and the situation was only partly redeemed by a rendering of the 'Gendarmes Duet' from *Genevieve de Brabant*, music by Offenbach!

The manner of my leaving BOC was strange. I had been asked by Chater to be ready to welcome two visitors from Cambridge, Professor A.H. Cottrell, FRS, Head of the Metallurgy Department and Dr J. Nutting. No special demonstration was required, we were merely to continue with the normal day's work, which was concerned with the earlier mentioned trials of argon through hollow electrodes using the Birlec arc furnace. The visitors duly watched a melt and, after casting the steel, I was introduced. I had met Jack Nutting before at BISRA (British Iron and Steel Research Association) conferences and had read that he had recently been appointed to the Chair at Leeds. I congratulated him and asked who his replacement was to be at Cambridge. He responded that, of course, he had no idea, not being involved in the appointment. In my sometimes impetuous way, not having any plans to move, I asked 'if I would do!' not expecting or getting any answer other than a chuckle all-round.

Later, after lunch, by which time I had changed into normal clothes, I was asked by Chater if I would take Professor Cottrell to his (Chater's) office and fix a taxi to take them back to the station. I protested somewhat since it seemed to me to be Chater's secretary's job and, in any case, the Underground was only a short walk away. Chater insisted and I accompanied Cottrell to the office. Once inside and alone together he asked me "were you serious about taking Nutting's place at Cambridge". My astonished response was that it had been intended as an off-the-cuff, humorous remark. He then said "well, think about it, and if you are interested, write to me and we will invite you up to Cambridge to have a look round". On asking my wife if she would like to move to Cambridge, her response was "what to do? There is no metallurgical industry there"; the thought that I might be an academic had not entered her head (nor mine up to that time). A visit to the Department of Metallurgy followed fairly rapidly and I was introduced to the, then, small staff. A formal offer from the University followed. It did not take us long subsequently to decide that we should take the opportunity, 'if only for a few years', and I have been there ever since!

My resignation from BOC caused considerable mayhem. I was told that there were plans to promote me into management, easy to say at that juncture, but of which I had no inkling. The only initiative on BOC's part up to that time had been to suggest that I should be Metallurgical Manager of a new joint company, Davy British Oxygen, set up to design and contract for oxygen steelmaking plants throughout the world, in competition with established firms with large design offices, using a very small outfit based in Harlow. I had rejected the offer since I could foresee failure quite rapidly, as, in fact, happened. I was also warned that at Cambridge I would 'disappear' professionally, and not be heard of again! However, I would not budge from my decision. The highest BORAD executive view apparently, although certainly not Bill Chater's, was that their premises had been visited by members of the University for the purposes of enticement of one of their staff. Although I was to learn, many years later, that Cottrell had already identified me as the man he wanted, my own forwardness had forestalled any enticement that they might have considered making. In any case, I had imagined that once determined, my move would have been welcomed and used to develop joint research projects with the University, but in fact my move was opposed at every point. The time at BOC had been enjoyable, but it was a pity that the last acts had to be as they were, with severe restrictions placed on me whilst working out my contract notice and no suggestion of future co-operation at that point.

I was very glad when, a few years later, a change at the top in BORAD resulted in a warm and friendly relationship being re-established, which has continued ever since with BOC. I was naturally very saddened to learn that this great British company has been taken over by overseas competitors.

Cambridge University, Department of Materials Science and Metallurgy, 1960 –

To live in Cambridge was not, in fact, a new experience for me. Both my mother and father were born in the town, although leaving in 1921, and their friends and relatives on my mother's side remained. Thus, many holidays in the 1930s were spent with them, particularly after my father was killed in 1932. During the intensive bombing of the London area during 1940–42, I was evacuated to Cambridge and attended the Cambridgeshire High School for Boys, one of the best grammar schools in the country. As a boy, I had explored every part of the county on my bicycle and even enjoyed early morning swims in the Granta from the Sheeps Green, Newnham, bathing sheds. It was thus with high spirits that we moved.

HISTORY OF THE DEPARTMENT AND SUCCESSIVE PROFESSORS

Having given a brief account of the origins of the Royal School of Mines and its impact on metallurgy, it is clearly necessary to balance that with similar treatment of the Department to which I moved in 1960 and where I have spent most of my working life.

It was in the Laboratory of Sidney Sussex College that it began, with the work of F.H. Neville and C.T. Heycock. Neville (1847–1915) graduated in Mathematics at Sidney in 1871 and in the same year became a Fellow. Heycock (1858–1931) graduated with a First class degree in Natural Sciences from Kings College in 1881, being elected to a fellowship in 1895, and between times establishing himself as a successful coach in science in partnership with E.H. Griffiths, a Fellow of Sidney. It was through Griffiths that Heycock was given the opportunity to carry out research in the Sidney Sussex Laboratory and the research partnership of Heycock and Neville began. Their first paper together in 1884 was non-metallurgical, on the molecular weight of ozone, but from then on they switched their attention to metallic alloys, with meticulous experimental work on solutions, with a classic paper in 1891 on alloys of gold and cadmium in tin, where, even today, the results are of some significance in the fabrication of circuit boards for semi-conductors. They were among the first to show the existence of stoichiometric compounds between metals, in the gold–cadmium system. With the assistance of Griffith, in developing platinum resistance pyrometers, they were able to extend their work, to higher temperatures than possible with the mercury thermometer used for the tin work, with a new standard of accuracy for temperature measurement at such levels, for example 629.5°C for the melting point of antimony as compared with the current value accepted more than 100 years later of 630.5°C, with a similar high quality of measurement for many other pure metals.

Towards the end of the 19th Century, there was a growing interest in the use of reflected light microscopy for the study of metal structure, following Sorby's studies of irons and steels in Sheffield during the 1860s. The technique was first employed at Cambridge by J.A. Ewing and W. Rosenhain in their studies of the deformation of metals. Both Sir Alfred Ewing and Rosenhain were at St John's, Ewing as a Fellow and Rosenhain as a 1851 Scholar. In 1899, they presented the Bakerian Lecture to the Royal Society on the subject. Heycock and Neville were clearly impressed and adopted photomicrography for their own work in the interpretation of alloy microstructures and incorporated the growing appreciation of the Gibb's Phase Rule in the preparation of the phase equilibrium diagrams that they were producing, for example gold–aluminium. Their 'star' contribution came in 1903, in the Bakerian Lecture to the Royal Society on the copper–tin system that contains 101 photomicrographs and a phase diagram of high accuracy, considering the techniques of the time.

For their work on gold–aluminium and other gold alloys, Heycock and Neville were provided with the precious metal on loan through the offices of Johnson and Matthey, gold refiners based in Hatton Garden, who also very generously carried out free assays on the alloys. Matthey had been Prime Warden of the Worshipful Company of Goldsmiths in 1872 and 1894 and was very influential on its Court. No doubt the close working relationship between the two men was instrumental in persuading the Goldsmiths to endow a Readership in Metallurgy for Heycock in 1908 and enable accommodation for him in the University Chemical Laboratory. At the same time, Neville retired and the Sidney Laboratory was closed, being out of use apart from storage. In 1910, a particularly spectacular Bump Supper bonfire was held in Cloister Court of the college and much of the Laboratory contents were burnt (the Bumps are annual rowing races between Cambridge colleges.)

"Heycock was an excellent lecturer" and "had few equals as a teacher in the laboratory; his deliberate method of working and his sarcastic denunciation of slovenliness inspired respect and awakened the spirit of emulation"; this from the obituary written for the Royal Society by the Head of the Department of Chemistry, Sir William Jackson Pope, a character well-known to my father. It is no surprise, therefore, that with such a reputation he was persuaded to give an evening lecture to the New Museums Club (assistant staff in sciences) on 'Metallic alloys' in the Cavendish Laboratory lecture theatre, which was subsequently reported in great detail in the *Cambridge Daily News* on 11 April 1911. One cannot imagine such a happening today, but it fits with the general ethos of self-help in education. There is little doubt that my father was present, as a prominent member of the club and later to be its President.

A new extension to the Chemistry Laboratory for metallurgical research, also funded by the Goldsmiths Company, was opened in October 1920, giving Heycock space that he desperately needed and it remained the main metallurgy laboratory from that time onwards, until Chemistry moved to new buildings in Lensfield Road. The next 'big' name in the metallurgical field in Cambridge, after Heycock, was U.R. Evans, who began his research in the Chemistry Department in 1922, with the work being brought under the control of the Metallurgy Department in 1945.

Ulick Richardson Evans, born in 1889, was a student at King's College, Cambridge (1907–11) taking Part II of the Natural Sciences Tripos in Chemistry. As a student, he has been supervised by Heycock and stimulated by him to an interest in metals and their electrochemistry in particular. After a short spell in Wiesbaden and then London, he joined the East Surrey Regiment in 1914 and served until 1919, mainly in the Middle East. After the war he returned to Cambridge, to facilities provided by Heycock in the new Goldsmiths' Laboratory and by Professor T.M. Lowry in the Department of Physical Chemistry. He lectured to both engineers and scientists from 1924 to 1954. This length of service being reflected in the fact that T.P Hoar (inevitably 'Sam") was one of his first research students and J.P Chilton one of his last. Eventually, in 1933 he was made an Assistant Director of Research and then Reader in 1945, followed by Fellowship of the Royal Society in 1949.

So many of the established theories of the electrochemical nature of corrosion, differential aeration, passivation by oxide films, cathodic protection, corrosion fatigue etc., originated with him that he has suitably been called "the father of the modern science of the corrosion and protection of metals." In 1960, aged 71, he published the 1000 page treatise *Corrosion and oxidation of metals*, with two supplementary volumes when aged 79 and 87, and continued to generate new ideas, not always concerned then with corrosion, until his death in 1980, aged 91. A truly remarkable man. Although I recognised his outstanding achievements and revered position when arriving in 1960, the fact that he was the living link to Heycock and the origins of the Department had not then occurred to me, for no doubt he could have given many interesting recollections of those early days that I could have recorded.

A familiar face in the corrosion laboratories for many years was J.E.O. Mayne (Jack). Given laboratory space, he was able to obtain funds and support not only his own research, but also, through his personal generosity, the work of the others in the field, both during his lifetime and subsequently. Having formed the firm Vinyl Products Ltd. and been responsible for the introduction of polymer-base paints, his interests were particularly in the field of paint films on metals and their protective role.

The Goldsmiths Chair in Metallurgy was endowed by the Goldsmiths Company in 1932 and Professor R.S. Hutton was appointed. After graduating in Chemistry from Manchester University, and with postgraduate experience there and on the Continent, Hutton had started his metallurgical career in the 100 year-old family business William Hutton Ltd (later to become the Sheffield Flatware Company) in 1908. With the impending failure of the business in 1920, he took the post of Director of the British Non-Ferrous Metals Research Association, until his move to Cambridge, building it up to be a very considerable force for the metallurgical industry of the country. Metallurgy was not taught as a final year subject in its own right, however, until 1937, and after the Second World War, as elsewhere, emphasis on chemical aspects gave way to physical metallurgy. Professor G. Wesley Austin, who succeeded Hutton in 1945, was to promote this change, particularly importantly in the mechanical properties of metallic materials and the relationship with microstructure, with the installation of a number of fatigue and other testing machines, and the commissioning of the Department's first electron microscope, a Siemens 60 kV model, much used by R.B. Nicholson when a research student and his supervisor Dr. J. Nutting, whose post I took when he went to the Chair at Leeds. Robin Nicholson was a Demonstrator at the time that I joined the department, soon to be a lecturer and then a Professor at Manchester in 1966. His subsequent careers at INCO (Managing Director), Chief Scientific Advisor at the Cabinet Office and then Pilkingtons (also MD), all confirmed the ability of this charming man. As Sir Robin, he became President of the Institute of Materials 1996–97.

I never knew Professor Hutton, but I had met Professor Wesley Austin on several occasions during my employment at British Oxygen. He was a small, neat man with a puckish grin and social charm. He had many contacts in industry and in government laboratories, often associated with his work with the Admiralty during the war and these contacts were particularly important in obtaining essential research equipment for what was a very poorly equipped and funded laboratory by the rapidly changing standards of that time. The expansion in provision was made possible by the increase in space made available by the move of Chemistry to Lensfield Road. Student numbers were also increasing during the 1950s, with recruitment into the Metallurgy course in Part I of the Natural Sciences Tripos and from Part I into Part II, responding to the quality of teaching associated with lecturers such as T.P. Hoar, J. Nutting, G.C. Smith and, once he had arrived, A.H. Cottrell. Recruitment of students onto courses at Cambridge is very much a competitive matter, with numbers responding quite quickly to a perceived quality of teaching, both in relation to direction of studies by the

Fig. 15 Department of Metallurgy, University of Cambridge, 1965. Staff members in the front row are: Piers Bowden, Mike Southon, Robin Nicholson, Tony Kelly, Brian Ralph, Gerry Smith, 'Sam' Hoar, Alan Cottrell, JAC, John Chilton, Jack Mayne and Graeme Davies. Alan Windle, present Head of Department, then a research student, stands behind Alan Cottrell. My three research students at the time(Charles Desforge, Tony Booth and Peter Maunder) form a group in the back row (Eaden Lilley, Cambridge).

student's College and, very effectively, to peer views. In Part I, particularly, the numbers of students attending lectures can be large and a confident, outgoing and enthusiastic style is an advantage. Knowledge of the subject with an enthusiasm to transmit that knowledge are essential ingredients in teaching at all levels.

After retirement, Wesley Austin continued to live in his 'baronial hall' near Thaxted, Essex, which was beyond the 10 mile limit normally imposed by the University for the residence of academic staff. It was characteristic of his dealings with University authorities that he took little notice of even the then relatively small amount of bureaucracy. He was very much a law unto himself, which did not always endear him to University officials. He and his wife visited us for tea soon after our move to Cambridge, when his sense of fun charmed our two young sons, particularly when, driving off, his car horn played a tune – very loudly!

Wesley Austin was succeeded by A.H. Cottrell in 1958, who remained in the Chair until 1965. Alan Cottrell was already established internationally as one of the leading names in academic metallurgy. He had been Professor of Physical Metallurgy at the University of Birmingham from 1943–49 and Deputy Head of the Metallurgy Division at AERE Harwell (1955–58), and later to become Sir

Fig. 16 B. Ralph, J.P. Chilton, A. Kelly, D. Dew-Hughes, R.B. Nicholson, A.H. Cottrell,
M. Southon, G.C. Smith, G.A. Chadwick, G.J. Davies, JAC.

Alan. Work in the physical metallurgy area, particularly electron and field ion microscopy was built up under Alan's direction, but he, like us, was very much aware of the need always to provide a balanced metallurgy course, containing a full range of natural science interests, particularly in the early stages of the Natural Sciences Tripos, so as to interest students with either preliminary physical or chemical interests up to that point, but ensuring that the material was markedly distinct from that appearing in either physics or chemistry courses. Thus, to a degree, my recruitment and the retention of a good corrosion content with the participation of T.P. Hoar and J.P. Chilton, and later G.T. Burstein and J. Little, might have attracted chemists. As indicated earlier, Sam Hoar was, in fact, made a Reader in the mid 1960s, the only one in the Department at that time.

Although Alan Cottrell's primary interests were teaching and research, he did not shy away from involvement with the central university authorities, although finding some of the committee work somewhat trying. He knew that as a relatively small department, it was important to participate and be aware of likely developments. In particular, we were vulnerable to a take-over by larger, more powerful departments. In fact, the Deer report to the University (published in the *University Reporter* 1965–66, p. 546) recommended that the work of the Department should be redistributed between Physics, Engineering and Chemical Engineering. Alan had marshalled all our efforts in defence of the Department and fortunately the recommendations of the report in relation to us were never implemented. The Department photograph, taken in 1965, is shown as Fig. 15 and the Examiners Lunch Party gathered in St John's, also in 1965, is shown in Fig 16. With the notice that Alan Cottrell was to leave soon after, to take up the post of Chief Scientific Advisor to the Ministry of Defence, we were naturally very apprehensive of the future. However, after an interregnum for the 1965

*Fig. 17 Department of Metallurgy, University of Cambridge, 1966, Professor Honeycombes'
first year. Geoff Chadwick now appears in the staff row. Tony Cowen and Tony Booth are in the
back row. (Eaden Lilley, Cambridge).*

Michaelmas and 1966 Lent Terms, with Gerry Smith as the excellent Acting
Head of Department, we were very pleased with the appointment of Professor
R.W.K. Honeycombe from Sheffield to the Goldsmiths Chair. The 1966 De-
partment photograph, with Professor Honeycombe in the centre (Fig. 17).

Robert Honeycombe's initial experience after graduation was with CSIRO, in
his birthplace, Australia. He came to Cambridge in 1948, firstly as an ICI Research Fellow in the Cavendish Laboratory and then as Royal Society Armourers
and Braziers Fellow. In 1951, he was appointed to a lectureship in Physical Metallurgy at Sheffield University, rising to Professor in 1955, before returning to
Cambridge and the Goldsmiths Chair in 1966, and was later to become Sir Robert.

I had first encountered Roberts' name when engaged in the bearing metal
research at J. Stone and interested in the thermal fatigue problems commonly
encountered in metals of tetragonal structure. He had co-operated with Walter
Boas in work on the same subject when at CSIRO, and their published results
were of considerable interest. At Sheffield, he had established strong relationships
with the local steel industry, indicating a real interest, other than an academic
one, in the application of research. The physical metallurgy of steels continued
to be his prime interest throughout his metallurgical career.

Roberts' contribution to the Department cannot be overestimated. He was
always friendly, accessible and prepared to listen to his staff, giving them the

In the 16th century, tripos was the name given to the three-legged stool on which the champion of the University sat when undergraduates were admitted to the degree of Bachelor of Arts. The term was then applied to his opening speech and later to the Latin verses of the new graduates. In the development of science teaching, the Cambridge Natural Sciences Tripos came to mean the course taken by an undergraduate where, in each of the first two years for Part 1, he or she would normally read several experimental subjects and mathematics, later moving to concentration on a single subject.

As recounted above, the Goldsmith's Company endowed a Readership (Heycock) in 1908, assisted with the building of a Laboratory in 1920 and then endowed a Professorship (Hutton) in 1932, establishing an important presence in the Department of Chemistry. Assisted by Dr D. Stockdale, Professor Hutton concentrated on developing the teaching facilities for the subject, which at that time was only examinable within the Chemistry papers.

In 1937, the University instituted a one-year course for a Tripos examination for third year students who had already passed a Part I examination in the second year, in the natural sciences or mechanical sciences, i.e. a Part II Metallurgy was established. In 1942, metallurgy was included as a possible half-subject (towards the three required) for Part I of the Natural Sciences, which could be taken either in the first year (Prelims) or the second year, thus giving the study of metallurgy proper recognition in both Part I and Part II (third year). By 1954, there were over 100 students in the Part I class and 30 in Part II, taught by a total staff of Professor Wesley Austin, Dr U.R. Evans, Dr T.P. Hoar, Mr G.C. Smith and Dr J. Nutting. Students for the half subject in Prelims or Part I would have three lectures and six hours practical work in Metallurgy per week. In Part II, there would be an average of nine lectures and fifteen hours of practical work per week. An interesting feature of the Department was, and has remained, the system of demonstrating, where practical classes are in the charge of a member of staff, but in addition, research students carry out up to six hours demonstrating per week. To ensure that students have satisfactory understanding of both lecture and practical material, a system of College-organised tuition in small classes operates, known as 'supervisions', with teaching either by academic staff or by senior researchers.

This organisation of the Tripos was still in place on my introduction to the staff in 1960. Over the years since major changes in the Tripos have taken place. The metallurgy half subject, in either the first or second years, was replaced by a joint first year (IA) course (with the Department of Mineralogy and Petrology) Crystalline State, and by a whole subject option in the second year, Metallurgy (Materials) IB. Someone eventually graduating in Materials Science and Metallurgy, having taken it as the sole subject in the third year, will usually have read these two options in the first two years, accompanied by a range of others to the equivalent of the required three in each year, but most commonly starting with Physics and Chemistry in IA. In many colleges the study of mathematics at a suitable level to accompany the experimental subjects has been a requirement for natural scientists in their first year, even if they have leanings towards the biological options. To cope with the expanding subject and, hopefully, to ensure adequate depth, recent changes have seen the introduction of a fourth year in Material Science and Metallurgy, i.e. a two year concentration on the subject in Parts II and III.

A great advantage of the Natural Sciences Tripos system is that it provides flexibility of choice for the student within Part I, and the possibility of changing direction without sacrifice at the end of both the first and second years. It is even possible to change the overall subject totally after Part I. An outstanding example of this is revealed in the academic career of Colin Renfrew (*see* p. 107) where his Part I Natural Sciences learning was followed by Part II Archaeology and where the application of science for the archaeologist was to be an important theme in his subsequent development. Others may change from Natural Science to Engineering or even to Law after Part I.

ARRIVAL, SEPTEMBER 1960 – IN AT THE DEEP END!

The staff in the Department of Metallurgy were welcoming and I was allocated the largest office I have ever had, before or since, and 'half' a secretary. The Department was housed in the old Chemical Laboratory building, fronting onto Pembroke Street, vacated when the chemists had moved to new premises in Lensfield Road. Considering the teaching load at the time, staff numbers were small, almost certainly a contributory factor to my spacious accommodation, initially at the rear of the building but then over the front entrance onto Pembroke Street. My staff colleagues were J.P. Chilton, T.P. Hoar, A. Kelly, R.B. Nicholson, G.C. Smith and of course, Professor A.H. Cottrell. No Readers had been appointed after the retirement of Dr U.R. Evans. With such a small staff the teaching load, covering the Metallurgy half subject in Prelims and Part I and the full subject in Part II, was heavy. Looking back to the *University Reporter*, student numbers taking the subject in the early 1960s were about 80 in both Prelims and Part I and 30–35 in Part II. Student numbers in Part II increased during the 1960s and 1970s to a peak in one year of about 60, but more usually between 40–50, a result I believe of the dedicated, primary attention of the staff to the quality of teaching, both in lectures and in being alongside the students during practical sessions. In 1962 there were only 26 research workers. Three or four research students per member of staff was usual and administration loads were relatively light.

In my memory, teaching was the pivot around which the Department revolved. By 1965, there had been staff expansion with B. Ralph (Brian), G.J. Davies

(Graeme), P. Bowden (Piers) and M.J. Southon (Mike) joining and amongst the research students was the present Head of the Department, Professor A.H. Windle (Alan) (Supervisor Gerry Smith) (Fig. 15). 1965 was, in fact, an eventful year all round, with Alan Cottrell leaving to become Scientific Adviser to the Ministry of Defence, to be replaced, as recorded earlier, by Professor Robert W.K. Honeycombe from Sheffield University (Fig. 17). Alan continued in the Civil Service until 1974, by which time, as Sir Alan, he was Chief Scientific Adviser to HM Government. Back at Cambridge, as Master of Jesus College, he was again a major player in University affairs. He is, without doubt, the greatest metallurgist of his day and it has been an honour to be associated with him. In retirement, he continues to produce fine theoretical, published work and, when possible, takes his place with the rest of us in the 'old timers' corner of the tea room! The expansion of the Department is illustrated by the photograph taken in 1990 (Fig. 18), the year of my 'official' retirement.

Joining the Department in October 1960, I was not able to take on any research students immediately, since all allocations had already been made and there had been no time for me to seek any financial support. It was a requirement that to be able to teach at Cambridge one had to possess a Cambridge MA, conferred on graduates from other Universities holding posts, under a special statute (B III 6). In order to take the degree, however, it was necessary to be accepted as a senior member of a College who would then put you forward for the degree. After discussion at a Staff Meeting it was suggested that an introduction should be made to St John's, since this was a large College without senior representation in the Department, from which, however, a considerable number of students read the subject in Part I and some in Part II. Gerry Smith, acting as Alan Cottrell's No. 2 and with wide contacts throughout the University, was delegated to make initial approaches to St John's. In due time, an invitation came to Gerry and myself to dine with the then President, Professor George Briggs, quickly followed by an invitation to both my wife and I to have lunch with the President and Mrs Briggs in his rooms in Chapel Court. All must have gone reasonably well, as I was duly accepted into St John's, taking my MA in January 1961 and commencing lectures during the Lent term, covering extractive metallurgy components of the Part II course.

College undergraduate teaching, i.e. supervisions, for St John's were initially quite a trial. In these I was required to teach on a two/three to one basis across all the Prelim, Part I and Part II courses, to include some aspects of crystallography and physical metallurgy, such as dislocation theory, which were new to me. Although neither were, in fact, members of St John's, my first Part II supervision groups included Richard Dolby, now Research Director of The Welding Institute, Abington and Alan Cooper, now retired but formerly a major player in the lead industry, mainly as Research Manager with Associated Lead, but latterly with Cooksons. My contact with both has continued on a shared research interest and friendship basis over the years.

Lectures could be carefully researched and presented, whereas the weekly face-to-face supervision encounters with bright students, endeavouring to sort out problems from any part of the course, was more demanding. As the years went by, greater specialisation occurred in the supervision system, perhaps inevitably with changes in the National Sciences Tripos and the introduction of subject options in Part II. I tried to base my approach to supervision on complete honesty with students and hard work, filling in the gaps revealed by shortcomings in response to them, and I hope I attained a suitable level of ability fairly quickly. I had the advantage of wide experience since graduation and had encountered in real life many of the manufacturing processes and applications of materials that were considered in the more theoretical aspects of the course. However, eventually the appointment into the Department of two Johnians, with initial Physics degrees, John Leake and then Ian Hutchings, enabled me to avoid any Crystalline State course supervisions. Over the years, many friendships were established with earlier undergraduate supervisees which have lasted and it is always a pleasure to renew contact whenever they are in Cambridge.

I recall my first Staff Meeting in Professor Alan Cottrell's office, seated with the others around the large mahogany table that had been left in the Department by Ulick Evans. Alan, introducing me, added that it had been possible to allocate an experienced technical assistant to me, able to weld and build furnaces. There was barely suppressed mirth around the table since I had been allocated an 'old-timer', only able to do relatively simple tasks and then in a slapdash manner and contributing little in terms of equipment design or research techniques. When the Department moved to a new building in 1971 and the new Process Laboratory was made my responsibility, the excellent Ron Chapman became my technical assistant. The laboratory had been designed for flexibilty in pilot plant scale operations, incorporating many of the lessons learnt at BOC, such as supplies on booms, an overhead crane and a heat resistant tile floor. Although it has seen some changes, it has proved to be a very valuable area for general use over the last 30 years. The direct supervision of the laboratory was eventually to be the responsibility of David Duke, who had joined the Department in 1965 as a lad of 16. He is now the much-respected Department's Chief Assistant.

The facilities in the Department, when wholly housed in the old Chemistry building on Pembroke Street, were totally unsuitable for the sort of research operations that had been established in the BOC Process Research Laboratory. There was some consideration of using again the outstation near the Cavendish Laboratory on Madingley Road that had previously housed the copper matte conversion studies carried out by Peter Young, who, after a spell in Australia was eventually to become Professor of Mineral Engineering at Leeds and my host on several visits as external examiner for graduate and postgraduate degrees. David Atterton, also a graduate student at that time, who, I think, had also used the Madingley Road facility, I was also to encounter on later occasions when he was Research Director of Foundry Services Ltd and in his continuing associations with the Department. Even when somewhat better areas were made available on the old Pitt Press site in Mill Lane and the melting facilities bequeathed by BOC on the closure of my old research foundry, with a venturi scrubber for fume removal, were established there, the necessary assistant staffing or experience for handling molten steel in worthwhile quantity were not there and fume escape caused aggravation to others nearby. Full realisation of the limitations on the type of process research that could be carried out was to be significant in relation to the design of the new Process Laboratory in 1971.

St John's College

Alan Cottrell had been, perhaps rightly, rather disappointed when I accepted the extra office of Junior (i.e. Domestic) Bursar at St John's College in 1963 at a salary of £500 per annum and a College Fellowship. Whilst I do not believe that it blunted my efforts in the Department for the four years that I continued in the post, it meant very long hours and exhaustion at the end. St John's is a big college and being responsible for most activities, apart from catering, that would be associated with a large hotel, occupying ancient buildings and much more, was a large extra load. When Contracts of Employment were introduced in 1967, and bureaucracy generally increased, I decided I could no longer continue. I had come to realise that I had to choose between a scientific or bursarial future and chose the former. I was, in fact, the last of the part-time Bursars at St John's, being replaced by a Mr A.C. Crook full time, who had been Director of University Estate Management Services. After a short interlude, I was Steward (i.e. Catering Supremo) of the College for one year, but then settled into the College post of Director of Studies in Natural Sciences (including, of course, Materials Science and Metallurgy), which I held for 15 years from 1972.

RESEARCH STUDENTS AND THE SELECTION OF RESEARCH TOPICS

Clearly the influences on my selection of research topics were my earlier background and the links with industry, the type of facility and staff available in the university context, the availability of financial support for the work (i.e. its attractiveness to funding sources) and, importantly, the attractiveness of the topics in a 'buyers' market for really good students. As far as internal students were concerned, the latter aspect would itself be influenced by the teaching they had received. Initially, with the substantial teaching of industrial metallurgy (e.g.

steelmaking, the formation and influence of non-metallic inclusions) and with a considerable technological content, my research students were often from our final year class, influenced perhaps, by enthusiasm for the subject in both lecture and supervisions. With a gradual imposed reduction of teaching in process metallurgy, the proportion of our graduates in the group investigating this area declined. Professor Derek Fray now runs a substantial research group in process metallurgy which mainly consists of graduates from other universities, often from overseas. Many of the students in my group were on CASE awards where some financial, material and facility support was arranged with industrial companies.

Perhaps, just because the type of project offered was usually directly relevant to the challenges of an industrial operation and, where new, sometimes difficult, experimental techniques had to be developed with no sense of continuing to turn an existing handle, the academic standard of the students was always high, as dictated in any case by the usually required level of a II:1 degree as a minimum for financial support. The group was never large, expanding to about six at any one time as the usual maximum. This was partly dictated by the number of students opting for the topics, but more by my own belief that this was the maximum that I could realistically supervise and give the level of support and involvement that I considered to be appropriate. No doubt more funds could have been hunted and wide advertisement employed to attract students and a bigger group built up, with post doctoral members taking some of the direct involvement. The students in the group were a great bunch and I always felt that a happy atmosphere prevailed. Some were naturally tidy and despaired of the others who were not; usually some individuals had special skills which they were willing to share, making a good team.

In the early days, research students would usually complete their work, write the thesis and possibly some papers within or just over the three year support period. As time went by, in spite of the availability of word processors, there was a gradual change, with almost four years before final submission not uncommon. I cannot begin to understand why this was. Maybe the expectations for a successful thesis, if only in their minds, had risen requiring more time. Perhaps, alternatively, they were more involved with outside interests and less dedicated to study.

In what follows, the research students and the area of work in which they were involved are briefly recalled in more or less chronological order, but with some modification to enable topic grouping.

RESEARCH ACTIVITIES

Non-metallic inclusions and the 'holey' ingot
When I commenced research with two students from the 1961 intake (P.S. Salmon-Cox and C.D. Desforges), the projects that were launched were related

to, but did not involve, large scale processing and, based on metallographic techniques, depended heavily on the new scanning electron probe microanalyser (Cambridge Instrument Co. Ltd., Microscan I) that had been committed to my charge.

Peter Salmon-Cox had been in the Part II class in the previous year and his remit was to investigate the origin and behaviour during solidification of the non-metallic inclusions in cast steel that were so harmful to properties. We were particularly interested in the large high melting-point aluminosilicates found in the bottom third of killed steel ingots, but more broadly in the general relationship between the composition of inclusions and their position in an ingot, as relating to the revealed macrostructure, indicating the stages of solidification. His work required a central slice of a substantial commercial ingot, where all additions and refractory materials employed in the production of the ingot would be sampled. Initial negotiations with Dr Saniter and Dr Jim Mackensie at United Steels at Swinden House, Rotherham were unfruitful. We were advised that what we proposed was not suitable for university research, much to my and Alan Cottrell's annoyance.

The Cambridge Instrument Company microprobe that we were to use had been originally developed at Tube Investments Research Laboratories (TI), Hinxton Hall by Peter Duncumb and David Melford after initial work in the Cavendish Laboratory. My involvement with TI, which is described elsewhere, had begun and led to contact with Park Gate Iron and Steel, Rotherham, then owned by TI, who agreed to help. A $3\frac{1}{2}$ ton killed steel ingot, uphill teemed, was earmarked from a cast and occupied one end of a large planing machine for several weeks, where it was machined down from both sides to produce a central slice $1\frac{1}{2}$ inch thick. The slice was sulphur printed and then cut into three sections for ease of handling. Even so, the pieces were heavy and required the construction of an etching bath from wood lined with coffin pitch! Using Oberhoffer's reagent the macrostructure was revealed, enabling specimens to be trepanned from positions chosen in relation to the various solidification zones thus made visible. By trepanning for the removal of specimens the integrity of the slice was maintained, enabling its continuing use for subsequent investigations. Peter Salmon-Cox worked for many years for United States Steel, but has now retired.

The origin and behaviour of non-metallic inclusions was, in fact, a very significant choice of investigation, leading to research not only by Salmon-Cox, but also by P.H. Maunder, A.R. Booth, Iku Uchiyama, J.C.M. Farrar, T.J. Baker, K. Gove, A. Segal and C. Honess. The original ingot slice provided material for many of the investigations and became known as the 'holey' ingot! Sadly, when my back was turned, it was included in a scrap metal collection from the department, although I believe the sulphur print (in two halves) is still in existence!

Droplet Oxidation

Charles D. Desforges came to me having graduated from Sheffield. The choice of topic was the oxidation of iron alloy droplets, an extension of the work that I had been doing at BORAD on dispersed liquid phase oxidation for silicon removal and on to spray steelmaking, although without help or encouragement at that stage from BOC. The study evolved into the use of an oxygen column, held in place by soap films, with inert gas above in the alloy rod melting zone and below in the quench. Various iron alloys were studied in this way. A particularly significant result was that at about 6%Si, the oxidation of a droplet was stifled. Microprobe examination revealed that this was clearly owing to the formation of a viscous silica film on the droplet surface, with no transport of oxygen to the molten iron beneath as with the iron silicate formed at lower silicon contents. This agreed with my experience at BOC in trials of decarburisation of transformer iron, where high silicon melts were largely unaffected by the injection of oxygen. Presumably, in that case, the oxygen bubbles were encapsulated by a silica film, preventing further oxygen uptake. Frequently amazing was Charles Desforges' photographic memory and his ability to recall whole pieces of text from papers. No doubt this advantage has served him well over the years of a significant career.

In 1962, I was joined by another of our Part II students, P.J.H. Maunder, studying the deformation behaviour of non-metallic inclusions in steel, considering the interface between the two and its effect.

Banded and Dual Phase Steels

In that year, there was an upsurge of interest within the Department into the subject of the fibre reinforcement of metals, encouraged by Alan Cottrell and with substantial activity by Tony Kelly and his students. The only contribution I could make was to suggest that an intercritical anneal and quench in a suitable steel with a banded microstructure could produce a 'fibrous' dispersion of strong martensite in ferrite. Another Part II student, R.L. Cairns (Bob), joined me and set to work. This research, and its relevance to dual phase steels for automotive body panels, has recently been reconsidered in the context of banded ferrous microstructures more generally. The fibrous martensite/austenite, produced much later by T.J. Baker at the RSM, was related to Cairns' work on the production of synthetic banded steels with martensite/ferrite structures by pack rolling different steels followed by heat treatment. The area of study also had relevance to the early history of steel, where pieces of iron were forged and folded to produce a banded structure with uneven carbon distribution maintained by segregation of solutes, e.g. phosphorus.

After leaving Cambridge, Bob joined the International Nickel Company at Suffern, New York as a research metallurgist. Later, he was moved back to the

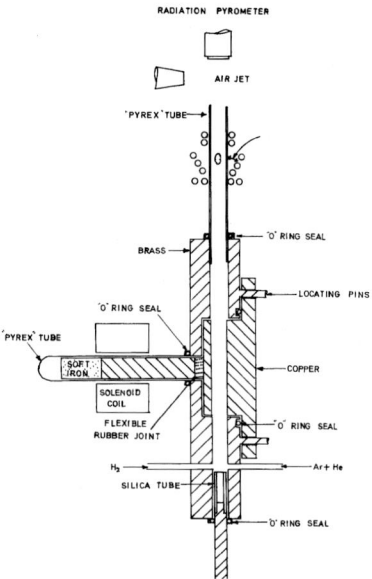

Fig. 19 Levitation melting and splat quenching rig for sulphides (Booth).

UK to work at the associated company, Henry Wiggin, Hereford, but died soon after in sad circumstances.

Sulphide Inclusions

In 1964, A.R. Booth (Tony) arrived from Sheffield to study the sulphide systems observed as non-metallic inclusions in steel. For much of his work, a levitated drop technique was employed, where an overall composition could be prepared by encapsulating sulphide ingredients in an iron bomb and melting in the HF levitation coil constructed for the purpose, at a temperature controlled by the power setting and the He/Ar mixtures surrounding the molten drop. The sulphide phase produced would equilibrate with iron in the molten state and was then splat quenched between copper blocks, fired together as the drop was released from the coil. Phase analysis followed. The development of the equipment was a particularly interesting challenge (Fig. 19). After several jobs in the USA, Tony is presently a director of Warner Lambert, concerned with R&D and with new business opportunities. We have occasional contact on his visits to the UK. At one point he was concerned with a novel way to produce thin plastic coatings on razor blades and I became involved to a limited extent through examining products.

The group was enlarged for a while by Iku Uchiyama, a university lecturer in Japan, who studied further the possibility of fluid silicate inclusions joining during heavy matrix deformation, arising from observations by earlier group members. Part of his approach involved the use of synthetic systems where filaments of silicate were drawn up into drilled steel blocks which were then rolled at controlled temperatures.

In 1966, J.C.M. Farrar (Chris) and A.G. Cowen (Tony) joined the research group. The former continued the work on non-metallic inclusions in steel, this time on the elongated inclusions and stringers that caused lamellar tearing in welds, particularly the products of silicon deoxidation in killed steels. The work was carried out in co-operation with The Welding Institute at Abington and was very well received on publication. Chris took a job at The Welding Institute after graduation and has remained in the welding field throughout his career, now Technical Director, Metrode Products Ltd.

Vacuum Deoxidation of Iron Melts

Tony Cowen has already been introduced as one of my stalwart colleagues at the BOC. I remember well his interview as a new RSM graduate with myself and Bill Chater. I had been told to move a certain item on Chaters' desk if I was happy to have him, which I was. Because of some hesitancy in Tony's responses Bill was uncertain, but thankfully took my indicated advice! After leaving BOC himself, he was employed by Telcon Metals, spending much of his time producing special magnet irons by the vacuum deoxidation of melts. It was an ideal topic for university research and, with the co-operation of his company, he moved to Cambridge with his family in 1966 for three years, as a SRC Senior Industrial Fellow and a member of St John's College. The results of the work, relating to carbon monoxide bubble nucleation, crucible materials and the minimum limits on carbon and oxygen contents imposed, were impressive and definitive and he duly obtained his PhD. It was good to be working with him again and I greatly admired the effort he put into his association with St John's, particularly in the sporting field. He is now retired and lives in Cornwall.

Electrolysis of Molten Slags

The two new research students that joined me in 1967 were both from the Part II class, but were very different characters. T.R. Shelley (Tom) quickly earned a reputation for producing Heath–Robinson-looking equipment which, however, worked. Components from a searchlight arc and crucibles from refractory-lined milk powder tins were used in the early experiments on the direct-arc electrolysis of molten slags. They were later to be transferred to the same 1 cwt Birlec arc furnace unit employed by me at BOC, which had been given, with other equipment, when the BORAD foundry was shut down, a gift which reflected the excellent relationship that had been re-established. His work on the electrolytic recovery of tin as oxide fume from both electrodes at a double faradic yield from

Fig. 20 Tom Shelley casting from the furnace.

molten tin-containing slags, with tin-containing cations and anions being dis-
charged and evaporated at the respective arcs, was absolutely excellent, but the
lash-up's of various heavy current DC supplies and the smoke produced by his
experiments, in spite of the electrostatic fume removal system, earned him a
reputation amongst assistants and some academics as the 'eccentric Professor'!
He was tall, somewhat gaunt and, with a shock of unruly hair, even looked the
part. His activities with the BIRLEC unit are illustrated in Fig. 20. These days he
is Scientific Editor of the magazine *Eureka*, most appropriately!

Continuing Non-metallic Inclusion Studies:
Inclusion Dispersions and Mechanical Properties
The other First class student from our Part II to join me in 1966 was T.J. Baker.
He took the non-metallic inclusion studies a stage further, establishing, in definitive work, a direct quantitative relationship between the dimensions of an
inclusion dispersion and the through-thickness properties of the steel. It was in
his work that we were able to translate earlier results that John Griffith and I had

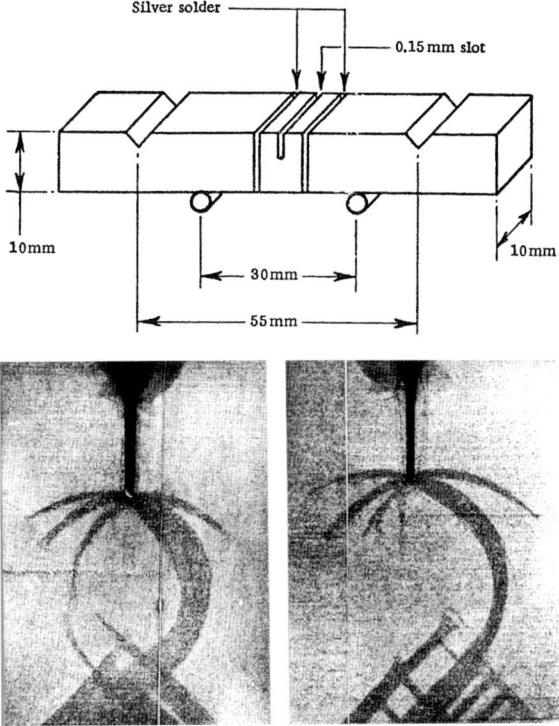

Fig. 21 Silver-soldering notched bend specimens in steel and nil effect on plastic hinge.

obtained for the joint constraint in soft-soldered brass to the employment of thin silver-soldered joints in steel, where the solder could be constrained to above the yield stress of the steel. This enabled Tim Baker to roll in our mill, from the maximum roll gape of 2 inch, specimens of a high sulphur steel in stages to the full extent reflecting commercial practice in the production of plate, resulting in varying elongation of sulphide inclusions. With the thin strip product he thus made, soldered between steel lugs and then notched in the thickness to the form of a toughness specimen (Fig. 21), he was able to establish the relationship between projected length of inclusions and crack opening displacement to through-thickness fracture. He was an outstanding research student who, after a spell at TI, Hinxton, subsequently joined the staff of the Metallurgy Department at the RSM, earning a reputation as an excellent teacher also. There was always supreme self-confidence in everything he did and he was almost always right, able to bring fundamental understanding to a practical situation. Such attributes suited him well as an expert witness in legal battles and he steadily built up a strong reputation as a consultant in this field, particularly in relation to fracture in the marine environment. The payment for such work is, of course, handsome, but legal cases can drag on and there was the inevitable conflict between his academic post and the much more lucrative form of consulting that he undertook. Court work is not everyone's 'cup-of-tea', but Tim was, and is, ideally suited to it and, after going onto a half-time contract at the RSM, he eventually felt morally bound to resign altogether, although retaining some teaching of the engineers at City and Guilds. If I had to call an expert witness in the physical metallurgy field there could be no-one better.

Kenneth Gove, who joined the group in 1969 from the Part II class, was probably the most productive research worker that I had the privilege to supervise. During his three years he built, from scratch, a high temperature microhardness tester, equipment for the pressure charging and the measurement of hydrogen in steel and completed his thesis and three published papers. His DIY approach was based on sound engineering ability; not for him were the inevitable delays incumbent in submitting work to the Department workshop, excellent though that facility was. A particular feature of the microhardness machine was that, unlike others, the indenter was heated to the required temperature as well as the specimen, greatly increasing the accuracy.

It was recognised, from earlier work in the group by Maunder, Baker and Farrar, that the deformation of non-metallic inclusions in rolled products, or their fracture to produce a stringer of broken particles, progressively reduced the through-thickness toughness. Ken's initial task was to establish the effect of temperature on the hardness of various phases in steel, particularly the non-metallic inclusions. The relative hardness or flow stress of an inclusion to the matrix in which it was set, determined the relative plasticity and deformation of that inclusion. Where there was zero hardness, as with a fluid silicate at high temperature, or a water 'bubble' in Plasticine, the inclusion could extend in the rolling direction to double the overall matrix deformation, owing to the concentration of stress at the hole in which the inclusion was set. Inclusion deformation ceased when it was twice as hard as the matrix. In the case of oxides the relative plasticity was zero at lower temperatures, with void formation, inclusion break-up and stringer development on rolling, with only a short temperature range in which any plastic behaviour occurred before the inclusion became liquid. In the case of sulphide inclusions the relative plasticity was high at room temperature, increasing further as strain ageing at about 300°C stiffened the ferrite matrix, but then decreasing progressively as the ferrite softened and recrystallised. With transformation of the matrix to harder austenite, the relative plasticity of the inclusion abruptly changed to a high value, then decreased again as the austenite softened with increasing temperature. This work was important in so far as it indicated the importance of controlling rolling temperature in minimising the deformation of sulphide inclusions and directionality of properties. In all cases, the fundamental information related to initial inclusion behaviour, as deduced from progressive measurement as deformation increased, since work-hardening of the matrix further complicated the results.

Working with J. Ward, for a relatively short period, the deformation behaviour of alumina clusters, as might be formed in aluminium-deoxidised steels, was studied using laboratory melts. In rolling, the clusters broke up with the formation of matrix voids and at the same time conical voids formed associated with individual separated particles. A definite relationship was established between the mean cluster length and the COD to through-thickness fracture, holding for rolling temperatures in the range 800–1200°C, a relationship very reminiscent of that discovered by Baker for sulphides.

With Ken Gove, it had been decided that with the high temperature hardness equipment available, capable of working in controlled atmosphere, the influence of hydrogen on the hardness of steel phases would be of interest, particularly in relation to the strain ageing process. The study yielded very interesting results. Further, with data on the characteristics of an inclusion population, this could be related to the hydrogen occlusivity of the steel, using the hydrogen charging and analysis equipment that he had built. S.L.I. Chan was subsequently to continue this aspect of the work. After leaving Cambridge, Ken tried to find a market for his microhardness machine and one or two further units were built. Sadly, none of the scientific instrument companies that he approached were interested. Whilst ahead of any competitor, the market would clearly be very limited and initial development costs were quite high. However, the basis of the design is there in his PhD thesis for other keen research workers to use and hopefully acknowledge his lead.

In 1972, Colin Honess joined the group from TI Hinxton Hall, taking the work on inclusion deformation and the effect on properties further on samples trepanned from the 'holey' ingot and applying ultrasonics and density measurement to the formation of voids associated with hard inclusions. At the time there was considerable interest in use of non-destructive inspection in relation to steel cleanliness, particularly if it could be on-line. Colin studied the behaviour of duplex and single phase manganese aluminium silicates during rolling deformation in the range 800–1000°C and the influence of the matrix structures on that behaviour, characterised by inclusion fracture and void formation with non-deforming inclusions. He studied the influence of morphology and matrix structure on the short transverse fracture toughness, using the same technique of silver-soldered specimens as described by Baker. Matrix structure was shown to have a controlling influence when final deformation was carried out in the duplex γ/α range, 800–850°C. Inclusion morphology, however, was established as controlling in material subsequently normalised at 900°C or by rolling in the single phase austenite range, above 850°C. A direct relationship was obtained between maximum inclusion aspect ratio and the crack opening displacement at maximum strength. The influence of void formation on the effective length of an inclusion was qualitatively demonstrated. In terms of the ultrasonic testing, the varying background reflection responses could be explained in terms of matrix structure and inclusion morphology, with a qualitative relationship between high ultrasonic background response levels and poor fracture toughness properties.

Observations from earlier work suggested that there was also an effect of size on the deformation of inclusions, with deformation absent at small size, and in 1972 Agnes Henderson (later Segal) also joined from the Part II class, having been one of my supervisees. Her task was to carry out a quantitative investigation of this size effect in sulphides. The most important result emerging from her work was to confirm, quantitatively, that the initial size of manganese sulphide inclusions does indeed influence their deformation behaviour, most significantly so at sizes below about 4 µm diameter. It was established, for example, that during rolling at 1000°C an MnS inclusion of 0.65 µm remains undeformed after the specimen in which it is set has undergone a 75% reduction. From this result, a value of 6.5 J m^{-2} for the energy of the MnS/austenite interface was deduced.

Agnes was an excellent research worker and after leaving Cambridge took a post-doctoral research fellowship in the department of Metallurgy at the University of Leeds with Jack Nutting. Moving with her husband to the West Country, her next post was with BAe at Filton, Bristol, as a development engineer. With a family there came a period when she had part-time teaching posts, then when back in London she worked for the Copper Development Association for six years. In 1996 she joined the Institute of Materials to become Manager, Membership Services in 1997. To our loss, and their gain, in 1998 she transferred her allegiances to the Institute of Electrical Engineers to become Professional Development Manager.

With financial support from his government, Fernando Bastian came over from Brazil with his family and quickly established himself in the group. With powder forged steels, non-metallic inclusions, principally oxides formed during atomisation to powder, are likely to reside at prior particle boundaries and form an easily-linked network, thus producing low toughness in the products. Higher oxygen contents decrease the spacing of the oxide inclusions and decrease the resistance to fracture. We were very fortunate to have the help of Gordon Brown, then at the GKN Group Technological Centre, in providing powder forged steels produced under varying conditions. Fernando showed that pre-forging deoxidation treatments could reduce the population of inclusions in powder-forged alloy steels and raise the toughness, with even relatively stable Mn and Cr containing oxides affected with strongly reducing atmospheres. A very good relationship was established with this mature student and we maintain regular contact. He is now a Professor at COPPE in Rio de Janeiro.

It was clear from earlier work, particularly that of Gove, that the interfacial energy between an inclusion and the matrix in which it was set would have an effect on the deformation characteristics of such a two-phase system. As a generalisation, it had been found that where a weakly bonded non-metallic inclusion in steel was twice as hard as the matrix in which it was set, then inclusion deformation did not occur and conical voids developed at the interface in relation to the strain directions. This was usually accompanied by eventual inclusion fracture and dispersal into the voids. To investigate the influence of interfacial energy two models were employed by S. Ashok: copper encapsulating beryllium copper rods and aluminium–manganese or aluminium–copper alloys. In the former, the interfacial energy could be affected by the casting temperature for the copper and the initial temperature of the rods. With 'coldshut' conditions there was no bonding between rods and matrix. With a high casting temperature for the copper the surface of the rods could be melted and a good bond obtained. The hardness of the artificial copper–beryllium 'inclusions' could then be altered by heat treatment. In the case of the aluminium alloys containing $AlMn_6$ or $CuAl_2$ the intermetallic compounds reflected a condition of strong bonding to the matrix, which could be determined and then, together with their hardness at the rolling temperature measured with Gove's apparatus, related to the relative plasticity during overall deformation. Second phase deformation in the aluminium alloys frequently continued even when the intermetallic compound "inclusion" was six times harder than the matrix. It is regrettable that the various aspects of this interesting work by S. Ashok were never published. It had produced data and ideas which would have been of considerable interest to the aluminium industry.

It was now clear that the emphasis of the steelmakers was on the production of clean steel. The significance of non-metallic inclusions in largely determining the mechanical properties of steels had been recognised and steelmaking practice developed to reduce their incidence, their size and to control their shape. That this was so was no doubt stimulated by all the work that had been done by people like Professor Roland Kiessling in Sweden and Brian Pickering and Terry Gladman in the UK and perhaps to an extent by ourselves. A strong personal relationship was established with Roland and he was very supportive of our work and, with Eva, extremely hospitable during visits to Sweden for conferences and discussions.

Hydrogen in Steel

Gove had built hydrogen pressure charging and analysis equipment with which he investigated the effect of hydrogen on strain ageing in ferrite and the effect of hydrogen on the stability of voids associated with inclusions on deformation. He showed that steel would take up hydrogen at a significant rate to saturation level in equilibrium with the hydrogen pressure applied, even at low temperature, whether or not the surface was free from other chemisorbed gases. However, the rate of leakage from a hydrogen storage cylinder at ambient temperature was shown to be very low, even when assuming that once hydrogen was in solution in the steel there was no further barrier for its escape to the atmosphere. As to the amount of hydrogen occluded by a steel it was clear that this depended primarily on the extent of suitable interfaces for it to occupy, e.g. inclusions, grain boundaries and carbide interfaces.

Shell Research Ltd. were particularly interested in the work on hydrogen and, in 1980, enabled us to take Sammy Lap Ip Chan, a graduate fresh from the RSM, for further work, with contact maintained through Dr David Morgan of Shell. A year later we were joined by Miguel Martinez-Madrid who was supported financially by the Mexican Petroleum Institute and the National Council of Science and Technology of Mexico.

The reason for the interest of the oil industry was an increasing incidence of sour gas containing H_2S in oilfields, which charges hydrogen into line pipe, causing embrittlement. For the work of both Sammy and Miguel, the NACE hydrogen charging method by corrosion was employed, using an H_2S saturated salt solution. Careful control of the iron/steel microstructure was effected so that the variation of occlusivity for hydrogen could be determined in relation to aspects of a microstructure, e.g. grain size, coherency of ferrite grains, pearlite coarseness and other transformation products. All this was a major effort, the results of which were properly presented in the scientific literature, principally by Sammy who continued work in the field after leaving Cambridge and eventually taking a post as Professor in Taiwan. In a recent letter, however, Sammy sadly informed me that his work on hydrogen had now ceased, largely resulting from a lack of interest by students in such fields as corrosion studies, preferring instead courses which lead into the electronics industry. It would seem that a reaction against conventional metallurgy in University courses is world-wide. Both Sammy and Miguel were great characters and widened the social horizons of the group in many ways.

In the context of the interest of the oil companies in the question of hydrogen take-up by steels as a function of microstructure, L.J.R. Cohen (Laura) joined the group from the Part II class, having also been one of my supervisees. Her task was to investigate the influence of hydrogen on the properties of duplex stainless steels. These steels were already being used in fairly benign environments and were candidates for more rigorous service in sour gas conditions. Different solubilities and diffusivities of hydrogen in austenite and ferrite and the tendency of the latter to deformation twinning, complicate the effect of hydrogen, with large-grained material more susceptible to damage than fine-grained material, and where a large volume fraction of small equiaxed austenite particles homogeneously distributed throughout a ferrite matrix is the most embrittle-ment-resistant. Laura carried out superb work in a difficult area and her abilities have been recognised during her subsequent employment at Zeneca. We keep in regular touch.

This involvement with duplex stainless steel and the effect of hydrogen on mechanical properties led to some research interaction with Dr J.E. King during 1988–91, when she was on the staff at Cambridge. Julia had been one of my undergraduate supervisees before studying for her PhD with John Knott. Her subsequent career, first as an academic at Nottingham and Cambridge and then as an industrial scientist at Rolls-Royce, where she is now Engineering and Tech-nology Director, Rolls-Royce Marine Power, has marked her as one of the most outstanding metallurgists of her generation, an undoubted 'star', now a CBE.

Julia's research student, T.J. Marrow (James), investigated the effect of hydro-gen on the fatigue properties of a duplex stainless steel (Zeron 100) containing 55% of austenite in a ferrite matrix. Fatigue crack growth rates were largely con-trolled by the interaction of the crack tip with ferrite/austenite grain boundaries. Increasing boundaries decreased fatigue crack growth rate. The interaction of the hydrogen embrittlement (as studied by Laura Cohen) and fatigue was defined.

Metal Matrix Composites

Whilst the work with Bob Cairns on martensite/ferrite composites had gone well, on reflection this may have been an area that subsequently I should have avoided, for my further participation proved to be a 'red herring'. There has always been a keen industrial interest in the increased use of magnesium and its alloys by reason of its low density. I knew from earlier industrial experience at J. Stone and Co., that a magnesium–lithium alloy containing approximately 11%Li would be cubic and highly ductile, unlike the hexagonal magnesium itself. It seemed to me that a combination of such a ductile matrix with strengthening fibres could be an exciting lightweight composite. The MMC work in the De-partment was, and is, under the direction of one of my one-time Johnian supervisees, Professor Bill Clyne. He agreed to take the matter forward with me, although my participation was largely assisting Bill and his student, C.M. War-wick (Marcus), another, later Johnian supervisee, in the preparation of the al-loys. The hazard of melting and alloying rightly partly belonged to the person who had suggested the idea! Sadly, however, the considerable strengthening by fibres that was achieved was unstable owing to the ready diffusion of vacancies in the matrix or at fibre/matrix interfaces as a function of the stress fields created. Marcus did an excellent piece of work and no doubt we all learnt a great deal from the exercise, if only never to count your chickens before they are hatched!

Nickel-Base Alloys

The effect of surface oxidation on the properties of powder forged iron and steels had already been examined by Fernando Bastian as an extension of the non-metallic inclusion study. With the advent of inert gas atomisation and processing, the possible advantages of a powder metallurgical route for the production of nickel-base superalloy turbine discs had been realised, the goal being the production of a homogeneous alloy with a relatively fine grain size and having high strength at intermediate temperatures. However, even with the achievement of low oxygen levels during processing, a combination of oxide and carbide persisted at prior particle boundaries. This PPB phenomenon was investigated with R.E. Waters (Robert) using the superalloy API, which had been argon-atomised. The examination of powder particle surfaces posed major problems and he had to employ a wide range of techniques in the work. He found that a thin oxide film, about 6 nm thick, existed on the prior powder particle surfaces, measured directly or as exposed by the fracture of as-hipped material. An interdendritic array of primary MC, secondary MC and $M_{23}C_6$ carbides were present both on the powder surface and the interior of the powder, together with M_3B_2 borides and some sulphur-containing particles. On heating the powder, almost all the carbon diffused to the powder surface, with widespread precipitation of secondary MC carbide platelets, in the absence of oxygen. As might be expected, there was evidence for the nucleation of MC on stable oxide at the PPBs. During hot isostatic pressing, large carbosulphides were formed and grew significantly.

Research had been instituted in the Department under Dr J.V. Wood (later Head of Department at Nottingham) on the atomisation and rapid quenching of nickel-base superalloys. On his leaving Cambridge, I was asked to continue work on the contract which had been given by the National Gas Turbine Establishment at Farnborough. This had involved building a metal spraying chamber and Rob Waters went on to try to develop the process, with complex inert gas (helium) quenching jets as well as the atomisation gas. With the size of the chamber, however, the product was essentially a fine 'splat', of elongated form, rather than powder, similar to the product of sample preparation from molten steel using the high speed rotating copper disc principle. Whilst the APKI 'powder' was never oxide-free, its form and the oxide distribution created some interest and the equipment was eventually shipped down to the National Gas Turbine Establishment at Farnborough into the care of Ted Restall, a well-known and respected figure in superalloy processing circles.

Matthew Witt, a Cambridge graduate like Rob, took up the challenge of investigating further the microstructure of as-atomised API, on as-compacted alloy and isothermally forged samples, paying particular attention to borides as little was, at that time, known of their constitution, behaviour during processing or their influence on properties. Much in relation to the latter aspect was shown to

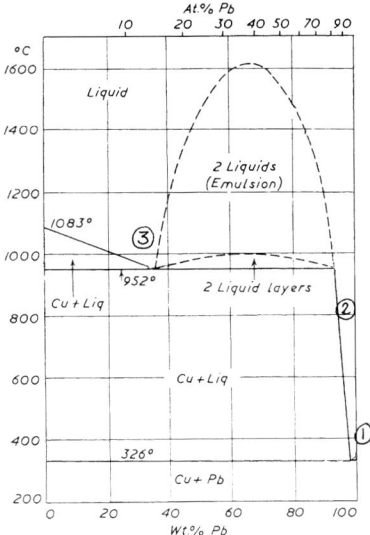

Fig. 22 The lead–copper system.(Metals Reference Book - Smithells).

depend on the aspect ratio of the particles where 'necklace' formation could be initiated during deformation. The linkage in my mind to the work on non-metallic inclusions in steel was clear.

The next worker in this field in the Process Metallurgy Group was S.J. Piggs (Susan) whose main effort was directed towards an investigation of faster cooling rates on the microstructure of API. Helium atomisation and melt spinning were employed to this end. After this work I did not feel that there was anything further that I could contribute in this field and no further members of the group were involved in superalloy processing.

The Lead–Copper System

Work on lead–copper alloys formed an important part of the Process Metallurgy Group activities over a number of years, partly as a result of associations with Imperial Smelting and partly Associated Lead (later Cooksons). As shown in Fig. 22, the system is a seemingly simple one of a monotectic and a eutectic. There are, however, complex considerations which arise in the following:

(1) the form and dispersion of the eutectic of Pb and Cu, which normally appears in an irregular or divorced form in cast low copper content alloys, as used in sheet or strip for buildings,

(2) the growth and separation of copper dendrites from molten lead during the copper drossing stage of lead refining,

(3) the solidification of the monotectic, where separation of liquid lead

from solid copper would need to be effected if high copper lead was to be refined.

These considerations are discussed individually in more detail below.

(1)   This was a most important study in the Group in terms of the length of time over which work was carried out and the number of students involved. Support was given by the Lead Development Association and member companies. Lead sheet was being produced in two ways; by the rolling of a cast billet or by a direct casting method. In the latter case, this was achieved with either batch production of thick 'antique' lead on sand beds for such applications as church roofs, or continuously cast by solidification onto a cooled drum dipping into a melt, with peeling off on the exit side, for more general building usage. Copper was always present to some extent in building lead as an impurity, but unfortunate experience with high purity lead, with gross creep after fixing, led to the establishment of a minimum level for copper being required in a standard to improve mechanical stability. Other than creep, the most common cause of failure then became thermal fatigue where expansion and contraction against fixing points could lead to deformation and cracking. The greater part of production at that time was milled lead where the original cast structure had been broken down and the copper phase dispersed by rolling. The morphology of the eutectic as a function of the original cast form (i.e cooling rate) was investigated and related to the dispersion in wrought products and the effect on mechanical properties and thermal fatigue. Directly cast lead had a dendritic cellular structure through the thickness of the sheet with the copper at cell boundaries. The free side showed some unevenness, although the gauge was generally remarkably consistent. Again, mechanical and thermal fatigue properties were investigated at different copper contents. Whilst giving a degree of mechanical strength and stability, the presence of copper does not improve the corrosion properties, particularly if coarsely present. The objective was to provide as fine a dispersion of copper particles as possible from both points of view. There was thus considerable attention to the breakdown and dispersion of copper in milled lead as a function of the rolling schedule.

The original research student in this field was Sue Vryenhof who worked extremely hard on the mechanical and thermal fatigue performance of the different types of lead available, introducing a large experimental programme both of laboratory and building site tests. Some of the external building tests, with strips fixed to a remote roof level wall so as to experience the maximum variation in temperature levels, are still in progress, although my ability to get there to assess results has decreased sharply in recent years. In spite of purposefully bad fixing practice, under the watchful eye of Dick Murdoch of the Lead Development Association, my last inspection did not reveal any cracks in the strips after 10 years. Sue had been a Part II Metallurgy student, but not one of my supervisees,

Fig. 23   Cross-section of a Roman minimus coin (×~17).

and joined the group a year or so after graduation. For my part, and I hope for hers, it was a very happy association and my wife and I regarded her very much at the time as a 'substitute daughter'. She now works for BNFL and has a home in Cumbria. The Lead Development Association group associated with the work was led by Alan Cooper (previously mentioned). The directly cast lead interest was represented, quite forcibly, by Koos Tuinenberg of Midland Lead. Other members included Roger Hill (Cooksons), Terry Boon (British Lead Mills) and B.N. Gearing (Broken Hill Association Smelters Pty. Ltd).

A particular problem for sheet lead roofing is underside corrosion and Helen Thompson, a graduate from Part II, worked on contract for a while investigating the corrosion of lead, the significance of copper particle dispersion and of differential aeration. This work has had some significance particularly as regards the influence of dispersed copper on corrosion in relation to work by the National Trust and others on historic buildings; attendance at discussion meetings meant renewed contact with Nigel Seeley, by then the head of the Conservation Department at the National Trust. Lihe Tan joined as a mature student from China where he had been a lecturer. His remit, in particular, was to consider the milling process in relation to the final distribution of the copper phase. He now works for EDAX out of Hong Kong as Asia, Pacific Regional Manager.

(2) As part of the refining process for molten lead, controlled cooling allows the separation of copper dendrites to form a dross floating on the metal surface. This dross is then separately treated for recovery of the copper. The amount of entrained lead will clearly be a function of the morphology of the copper dendrites (interarm spacing), the degree of wetting by lead on the copper and the degree to which the dendrites are broken up in the molten lead prior to forming the dross.

(3) If the copper content is high then the system moves into the monotectic with the formation of two liquids, with the lead-rich then precipitating copper dendrites as it cools from the monotectic temperature down to the eutectic and the other generating metallic copper and the lead-rich liquid at the monotectic.

The first time that I observed the products of the monotectic reaction was in the examination of a Roman minimus provided by Professor Ted Buttrey, through an association at the Fitzwilliam Museum. This tiny coin had, in fact, been produced by melting a roughly 50/50 mixture of copper and lead in a depression in a clay tile. The free molten surface gave one flat surface. On the other, lower side, a lead 'plug' separating in the two liquid region had been forced to the centre by the monotectic copper/lead growth extending from the free surface, which solidified first. Free dendrites of copper had subsequently formed within the still molten lead plug as it cooled down toward the eutectic (Fig. 23). A previous view had been that these coins had been drilled and a lead core inserted in order to provide a sensible weight, but it was clear that it was, in fact, the result of normal solidification of the somewhat unusual alloy system. The slight protuberance of lead from the lower face resulted from attack by lead oxide on the siliceous material of the tile mould surface below.

In the context of modern process metallurgy, there was the possibility of operating a lead-producing smelter such as the zinc blast furnace on a higher copper content charge and solidifying the copper metal from the monotectic on tapping from the furnace and prior to normal dressing. A thorough consideration of monotectic reactions and the influence of variables on the morphology of the copper/lead product was undertaken and it was clear that without the use of external force, it was going to be difficult to separate the phases significantly. Two practical approaches were developed, the solidification of the copper from the monotectic onto a spinning cooled rod, in order to throw off the molten lead phase before it became entrained, and pressing at temperatures below the monotectic in order to squeeze out separated lead. The student involved in work on considerations (2) and (3) was Stuart Sarson. The theoretical study of monotectics went well, resulting in the publication of a major literature survey, and there was some promise from the experimental work, which however, was not taken further or published beyond his thesis, and with the closure of research at Imperial Smelting the work ceased.

Soldering

With earlier experience of tin-base alloys, the strength of soldered joints and then more recently lead, including the context of thermal fatigue, it was natural that an interest would develop in the behaviour of fine soldered joints on circuit boards, where thermal cycling during operation can cause considerable problems. A joint SERC-CASE project was established with Plessey, then GEC Marconi, represented by Dr David Pedder, who had been a research student with Gerry Smith some years before. The research student recruited for the work was N.R. Green. Nick had been a student at City Polytechnic and like Professor H.K.D.H. Bhadeshia, one of my colleagues well known for his work on transformations in steel, had graduated from there in Metallurgy with First Class Honours.

The miniaturisation of silicon integrated circuits has been achieved by improvement of pattern transfer techniques and better device structure design, with a high density of inputs and outputs where lead bonding has to be achieved. In a flip-chip microsolder device the track terminations of the chip are overlaid by an array of solder-wettable pads, on which solder has been deposited to form discrete bumps over the pads. After inverting onto the substrate, which possesses a corresponding array of wettable pads, the whole assembly is heated to above the liquidus of the solder and bonds are formed.

The study focused on the characteristics of microsolder bond solidification, microstructure and reliability. At their very small size (30,000 to 80,000 μm^3) and spacing, the stress conditions developing at individual bumps in the array and the significance of microstructure was likely to be different to those at a larger scale, with differing sensitivity to thermal fatigue.

The microstructure of the eutectic Pb–Sn bonds were, in fact, not at all similar to those published in previous structural studies of the alloy, owing to the undercooling, the metastable conditions and the purity of the elements used in the bonds, with no pre-existing nucleating compounds. The smaller diameter bonds contained large lead dendrites, whilst the larger bonds had microstructures more like coarsened irregular eutectics. The matrix of the bonds were considerably enriched in tin. Also $AuSn_4$ intermetallic was found within the bonds, developed by the solution of gold from the underlying pads. Long term storage and/or thermal cycling resulted in this compound transforming to a ternary compound $Au(Cu\,Sn)_4$, nearer equilibrium, with copper substituting for tin.

Observations of the thermal fatigue degradation on thermal cycling were very interesting. In highly strained bonds, cracks were initiated after rapid coarsening of the lead phase, with the relatively few Sn–Sn and Sn–Pb boundaries or solidification defects playing major roles. The time to initiation of cracking decreased as a function of the coarseness of the microstructure. Once initiated, the cracks proceeded along Pb–Sn, Sn–$AuSn_4$ or Sn–Sn boundaries. Where no boundary existed, then the crack grew through the tin dendrites. The observed improvement of fatigue life, resulting from refinement of the microstructure, could have been due to the bonds deforming superplastically during the early stages of thermal cycling. The anisotropy of the thermal expansion coefficient of the tin phase in the bonds and the associated thermal ratcheting between adjacent tin grains and bonds was an important feature of the thermal fatigue behaviour and was also investigated using constrained solder strips, here the solder was rapidly solidified on thermally stable metallised Invar substrates and then thermally cycled.

Whilst the work on the solidification and microstructure of Pb–Sn solder bumps was duly published in *Materials Science and Technology*, the second phase on the thermal fatigue behaviour regrettably was not. Nick left us for the Castings Centre, IRC in Materials at Birmingham University and from thence, as a castings modeller, to VAW Motorcast Ltd, having done work for them whilst at the Centre.

The behaviour of solder bump arrays and other joint systems in integrated circuits could conceivably be modelled for performance prediction and this was an aspect taken up by my colleague, Dr E.R. Wallach, and his student, Paul Winter, in collaboration with BNR Europe. After my retirement it was kindly agreed by Rob Wallach that any bench space that I might need could be in his laboratory. In view of the work carried out with Nick Green, I was able to take some small part in these further investigations.

The Use of Plasmas in Process Metallurgy

A great colleague during my early years at Cambridge was G.J. Davies (Graeme), who later went on to much higher things as successively Professor in Sheffield, Vice-Chancellor of Liverpool University, Head of HEFC(E), and then Vice Chancellor of Glasgow University. He investigated the heating of mineral particles in flight in an RF induction plasma, using molecular spectroscopy to identify any dissociation which might be occurring. It was clear that in such a system, the particles were not entering the dense plasma regions and thus not reaching very high temperatures. I had taken a close interest in the work in view of the experience that I had had at BOC with powder additions made to electric arc furnaces via hollow electrodes using argon gas. When M. Anthony of BP Minerals International Ltd. expressed a wish for further work to be carried out on the use of plasmas in process metallurgy, with financial support, there was the opportunity to re-examine the approach. After discussions with various experts, it was decided not to pursue the RF induction or the usual plasma furnace approach but to try to utilise the 50 kW horizontal coalesced arc plasma system developed by J.E. Harry and L. Hobson at Loughborough University. The plasma gas from this type of 12 electrode plasma generator could be led into a sealed reactor, making treatment in a fluidised bed a possibility. Much effort was expended by M.C.L. Patterson (Mark), a graduate student from the Camborne School of Mines put onto the project, on the development of the system, e.g. synchronous pneumatic electrode movement for initiation and expansion of the plasma. Dissociation of minerals injected into the plasma was achieved, but operation as a proper fluidised bed required a gas volume which reduced the maximum temperature achievable to 1400°C, even at maximum power. This was below the temperature achieved in an equivalent plasma jet furnace. With a limit on the power available there were only a small number of options for improvement.

An excellent Korean student, Young Wan Cho, took the use of the fluidised bed reactor further, with a major redesign of the coalesced arc system, reverting to mechanical electrode control through stepping motors and generator jacket water cooling, electrode cooling and gas sealing. Further, a smaller reactor bed made of graphite was produced. By this time, however, the required focus had moved away from mineral dissociation to the exploitation of the reactor for the production of nitrides using a pure nitrogen plasma. However, even at maximum power, argon plasmas containing even small amounts of nitrogen could only be maintained for a few minutes and much greater power availability was required. Thus, with the limitations of the equipment available and the time already expended, it was decided to concentrate attention on the preparation of nitrogen ceramic powders using more conventional equipment. Nevertheless, it was still felt that the coalesced arc plasma could have been developed further technically with extra power.

Turning to normal heating methods Young Wan obtained very interesting and important results in the production of aluminium nitride where it was discovered that the overall reaction rate between aluminium and carbon in nitrogen to produce AlN can be considerably increased by using $Al(OH)_3$–C mixtures instead of αAl_2O_3–C. AlON was formed as an intermediate compound. Since, also, very pure, fine $Al(OH)_3$ powders are commercially available at lower cost than α–Al_2O_3 as starting grade, a new preparation route using $Al(OH)_3$ as starting material should provide economic advantages. An important point was that the morphology of $Al(OH)_3$ powders did not change during the reaction to AlN, making it possible to produce the latter at very fine particle size and narrow size distribution. The process was patented by the Carborundum Company (formerly Sohio Engineered Materials) of Niagara Falls who had taken over the BP interests in the matter. We understood that the route was going to be used but have heard nothing in recent years.

Young Wan also studied the production of silicon nitride and sialon. B′ Sialon powders were produced by reacting kaolin and carbon black in nitrogen at about 1450°C. A totally new slant was the influence of carbon structure, where a high structure carbon black in the starting mixtures facilitated the reaction, again a result of importance in the industrial production of the material by the carbothermal route. All this and the submission of award-winning papers was completed in the time that Y.W. Cho was at Cambridge, reflecting the tireless dedication to his work that he gave.

Other High Temperature Process Studies

Two projects were undertaken jointly with Derek Fray in the late 1980s. David M. Hudson used both RF induction and plasma heating in a study of the evaporative refining of metals, with concentration of activity in relation to the treatment of aluminium and iron scrap, with the removal of impurities as fume. One of Derek's primary interests has been fused salt electrolysis and the second joint study, with P.M. Copham (Piers) as the postgraduate student, concerned the use of rotating electrodes in fused salt electrowinning. Electrolysis cells operating with molten salts normally have a high energy consumption and the output of metal per unit volume, or space/time yield, is low as a result of the large interelectrode spacing employed. The need, therefore, was to reduce the electrode separation, but this required rapid removal of the electrolysis products to avoid recombination. Piers investigated the possibility of doing this at close spacing by applying a centrifugal force across the electrodes, using rotation, on the system $ZnCl_2$–KCl–NaCl. Although the current efficiency at small interelectrode spacing was substantially increased by rotation, gas escape from between the electrodes was a significant problem, although one which should have been capable of solution had time allowed.

ACADEMIC RESEARCH BASED ON ASSOCIATION WITH INDUSTRIAL COMPANIES

Association With Tube Investments Research Laboratories

Association with Hinxton Hall as a consultant had the welcome consequence of opening the way for young members of TI staff to join me on secondment to carry out research projects, registered for the PhD degree. The first of these (1969–72) was an attempt with James Tolley to understand the way in which 1% carbon–chrome steel in tube form, used for rolling bearing components, behaved on machining, particularly with cemented carbide cutting tools. Hopefully, the effect of existing production variables could be explained and recommendations made for improvement. After careful experimental study in the context of theoretical analysis of the machining process, it was confirmed that when using high speed steel cutting tools, increasing spheroidal carbide size and cold reduction increased machinability, while stress relieving or the formation of a pearlitic microstructure reduced it. Failure of high speed tools was by a combination of excessive flank wear at the outside of the cut and severe crater wear on the rake face, with abrasion being the main factor, but with a contribution from critical strain for shear failure.

With cemented carbide cutting tools, there was only failure by extensive flank wear controlled by critical strain for shear and there was a reduced sensitivity to the microstructure of the workpiece. In fact, small spheroidal carbides were preferred to a coarse distribution, with important prior heat treatment advantages. As before, cold reduction improved machinability; subsequent stress relief reduced it, partly as the result of increased tool/workpiece chatter. Tool crater wear was insignificant, largely owing to the presence of protective non-metallic deposits. At the higher tool surface temperatures of carbide cutting, oxidation of silicon from the metal combined with any non-metallic inclusions that might be present gave a surface lubrication effect, altering the chip form beneficially. The advantages thus shown for carbide tool use were of obvious commercial value and there was an undertaking to restrict access to Jim Tolley's thesis for a period. This also had some effect on the published account of the work, where only the results relating to microstructure control when using high speed steel tools was discussed.

The second secondment from Hinxton, in 1970, was Norman Fletcher, following up an idea on the use of fluidised beds of solids as quenchants, particularly for the heat treatment of powder metal parts in alloy steels where mechanical properties can be improved by the use of a post-sintering heat treatment procedure. The use of conventional water or oil-based quenchants can result in distortion, since there are very high values of surface heat transfer over a particular range of high temperatures, establishing a steep temperature gradient when there is a relatively low yield stress. To minimise distortion a quenchant has to be less sensitive to surface temperature in terms of the heat transfer coefficient.

Norman showed experimentally that heat transfer values approaching those of an oil quench could be achieved in a bed of millscale fluidised with air. Using a 0.3 m diameter bed, wall effects on bed form were minimised. The value of the heat transfer coefficient, or severity of quench, could be altered by changing the size of particles in the bed or by increasing the air flow rate through the bed (i.e. expanding the bed). The larger the millscale particles used, the lower the maximum heat transfer coefficient obtained, the less the sensitivity of this coefficient to air flow rate and the larger the air flow rate needed to fluidise the bed. The use of smaller particles gave a higher maximum heat transfer coefficient (greater exchange surface) and increased sensitivity to air flow rate and, of course, a reduced flow rate for fluidisation. Elutriation from the bed does, of course, limit the minimum size that is practicable.

As with oil and water, the heat transfer coefficient was found to be dependant on the surface temperature of the component being treated, but less sensitively so and no maximum was exhibited at intermediate temperatures. Therefore, the tendency to set up high thermal gradients was reduced and thus the tendency of the components to distort, which is an extremely important advantage of the fluidised bed. Just as in liquid quenchants, there was anisotropy of heat transfer when comparing horizontal surfaces with vertical surfaces where the former may be as much as 80% lower, resulting in this case from defluidised 'hoods' above the surface and gas pockets. Secondary agitation could be required to minimise this difficulty. The use of a fluidised bed as a quenchment gave a clean, dry, product and would be particularly attractive in a continuous process context. Whilst the idea may not have been entirely novel, its development into a useable bed for the continuous treatment of sintered alloy steel powder parts was a useful contribution by Norman, which should have been taken further. Sadly, the work was never written up as a thesis by him nor as a scientific paper after he had returned to Hinxton. A third secondment occurred in the person of Colin Honess. His work on non-metallic inclusions is described elsewhere. The co-operation of TI in these projects was greatly valued.

Association With Imperial Smelting Processes Ltd. (RTZ/RT)

As a consultant to Imperial Smelting Processes Ltd. (ISP), it was natural that topics would arise which were suitable for university research. After initial skirmishes into the consideration of the freeze purification of zinc, where problems arose from the cellular nature of the solid/liquid interface, the first of these topics was concerned with the protection of steel from attack by liquid zinc, of obvious significance in the handling of the metal by pumps and in holding tanks. G.D.S. Price (Geoff), a Part II student, took on the work, studying the mechanism of the attack as a function of temperature and of steel composition and developing a form of calorising to produce an adherent aluminium oxide coating, which gave encouraging results.

Whilst separate zinc and sulphide concentrates arise from ores, they together form part of the charge for the Imperial Smelting Furnace process as either sinter or roasted to oxide as pellets or briquettes. D.L. Canham (David) worked on the permeability problems arising from the formation of fusible lead phases (sulphates) in pellets. Instead of producing uniform mixtures for pellets, using two pelletising stages he put a zinc sulphide on a lead sulphide core, greatly improving permeability during reaction to the oxide product.

One form of charge to the zinc blast furnace could be briquettes, particularly useful for consolidating oxidic zinc powders from whatever source, including zinc and lead-containing drosses, into lumps suitable as part of the furnace burden. Hot, self-sintering bonding or cold briquetting with an additional binder could be used. With conventional briquetting rolls, however, the pressure exerted on the powder in the roll pockets is not uniform and, as a result, briquettes are liable to be weak and crack. A.R. Begg (Alan) developed a half pear-shaped pocket which gave a more uniform briquette density, and which was stronger in handling. To obtain high strength, the briquetting process should be hot, where diffusion in the zinc oxide contributes to strength and where there was shown to be an important effect of impurities on the diffusion process. This work went to pilot plant trials, run by Dr Chris Cross at Avonmouth, but was not taken any further for economic reasons. Injecting fine wastes into the zinc blast furnace via the tuyeres has proved to be a good way of dealing with such materials.

Alan was an impressive research student and his subsequent career was at TI research laboratories, followed by a move to BP and then to T&N, where he was formerly Managing Director of T&N Technology at Cawston and is now Vice President, Technology for Federal-Mogul (who acquired the T&N Group in

1998) in the USA. E.R. Wright spent three cheerful years investigating the absorption of metal ions by lignites, with varying degrees of success, but he underestimated the value of the interesting features that he had observed.

Fused Salt/Molten Lead Heat Exchange

A particular disappointment to me, however, was the failure to bring to fruition the work of R.F. Cochrane (Bob). At present, the liquid lead in the launders leaving the splash condenser of the ISF, is cooled in a system which, to ensure freedom from attack by separating zinc, only gives hot water as a product. If a cooling system could be devised that produced high temperature steam, there should be an increase in thermal efficiency. Our idea was to mix a fused salt at a lower temperature with the lead at the higher temperature in a vaned mixer device to a uniform mixture temperature and then to separate the two phases in the equivalent of a hydrocyclone. The fused salt could then be used in a steel heat exchanger to generate steam. After some small-scale laboratory experiments to define a suitable salt composition and operating variables, work commenced on a small pilot plant. The problems of circulating molten lead and molten salt, with the failure of heating tapes and the frequent leaks of liquids at high temperatures, were difficult for us to overcome. It was thus decided to transfer the plant to ISP at Avonmouth, where the intention was to use the greater practical resources and expertise there to carry out a critical assessment of the idea. Sadly, however, this did not happen. The equipment was set on one side and not long after the Research Department at ISP was closed down, by which time I believe corrosion had taken its toll. The idea of using a fused salt heat exchanger had become less interesting because of the installation of boiler panels at the Hachinohe ISF in Japan. In the end, there seemed a lack of economic incentive for power generation from low temperature heat sources.

The work carried out by Stuart Sarson on the copper–lead monotectic reaction was also promoted by our contacts with ISP, but is considered separately.

Apologies

At this juncture, I should confess that my research students were sometimes at a disadvantage in that, apart from the study of non-metallic inclusions in steel, the topics that introduced were often 'one-off', associated with a problem encountered during consulting at, say, ISP or an inspiration from remembered earlier industrial experience. Students tackling these topics did not have the advantage of an established group activity, where one student would take up where another had left off and where there was a bank of existing technical 'know-how' and equipment.

As already intimated, I am also conscious of the fact that not all the work carried out in my research group was written up in the form of scientific papers for a wider audience than the PhD thesis. This, I very much regret and I have always done my best to ensure that proper publication followed. However, sometimes completion of the thesis in the time allotted in relation to financial support is a close-run thing and once the student has left it is extremely difficult to insist that papers should be written. Unless there is very real motivation, time is not found when in competition with the pressures of a new job. On one or two occasions I have completed papers on my own, but since the students have been so closely involved, they have a right to be named authors and to do all the work for them would be to create a precedent that could be unwise. Part of the motivation for this present text has been to record all these efforts even if briefly, so that reference to the original thesis could follow.

SELECTION AND USE OF ENGINEERING MATERIALS

From student supervision in the early 1960s there was clearly a need for a unifying course at the end of the Part II final year, which would bring together factual information from earlier components of the syllabus in the context of the use of real materials in the fabrication of a component. With considerable emphasis on corrosion processes in the course at that time, the selection lectures given by Dr J.P. Chilton, a good friend throughout the Cambridge days, gave considerable emphasis to this critical aspect of material selection. I took over the course in 1967 and continued until 1988 and it broadened substantially in relation to my own experience and interest, expanding from eight lectures to eventually twelve.

It may seem anomalous that someone ostensibly responsible for process metallurgy teaching should end up with a major course of lectures, for over 20 years, on the selection and use of engineering materials. The explanation, moreover, is difficult to recall, but in truth my background was a very varied one to include teaching metallurgy to engineers in-house at BOC and at evening classes and, also, whilst at BOC, I was frequently required to investigate component or plant failures and to give advice on materials. One particular investigation I recall concerned cracking at the necks of Sparklets soda water siphon bulbs, attributable eventually to elongated non-metallic inclusions (sulphides) in the heavily worked steel at the neck. Furthermore, once at Cambridge, I was becoming increasingly involved with engineers as a consultant and very aware of the significance of an understanding of materials properties and their selection in the context of design and fabrication.

When I was a student, the teaching of materials selection, such as there was, involved little more than the recitation of specifications, compositions and properties, with little comment as to relationships with design or areas of use, regrettably a rather boring exercise. The approach that I took was that, after introductory lectures on general principles, the basis of cost etc., the greater part of the course involved dismantling various artefacts into component parts and discussing the selected material for each and how that selection and its fabrication fitted into the overall design. Components were selected to illustrate the use of a large range of materials. The companies that made the artefacts had previously been visited and their design and selection philosophy discussed with them. This interaction was a vital preparation for me.

In the early days, the course was given in the Babbage lecture theatre to a large audience of metallurgists, materials scientists, mechanical engineers and chemical engineers, with extensive use of television, enabling them to watch the dismantling process and see greatly enlarged projected images of components on a screen as they were discussed. Initially the cameras were 'choreographed' to a script, with Peter Rothwell, a younger member of staff, acting as production manager and cameraman, but I soon preferred to have more flexible arrangements, with the theatre assistant, Mr DeCastro, responding with camera to spoken requests.

It was clearly difficult to give such a course with depth to students from such a wide background. Professor M.F. Ashby (Mike) who had been elected to a Chair in Engineering, attended the course during one session. Mike had been a student and postgraduate in Metallurgy before going across to Engineering and had the intention of developing a major course on materials engineering in his Department, with another Cambridge metallurgy graduate, Dr Dai Jones. With the mechanical engineers eventually withdrawing for their own new course in the Department of Engineering, it was also considered best if the chemical engineers also had their own, concentrating again on corrosion behaviour in materials, with John Chilton again giving the course, to be followed later by colleague Dr John Little.

It was then possible to focus on the metallurgists and materials scientists with greater detail and with a better chance that they had at least heard of the materials under discussion! With reduced numbers it was then no longer necessary to use the large Babbage theatre and eventually the lectures were given in the seminar room in the Department, where the closer proximity of lecturer to students meant that television did not have to be used and components could be handed round or examined at the end of the lecture. The use of television and a large projection screen had also had unexpected consequences. In the case of one young man, the large image of a dental root drill, with demonstration of the cutting principle and the drill flexibility, was more than his stomach could take!

The obvious success of the approach taken encouraged me to think of writing a textbook organised in a similar fashion. David West at Imperial College was aware of my activities in putting pen to paper and informed me that a colleague of his at the RSM, Dr F.A.A. Crane (Andy), who gave lectures on selection, had started to write as well and that it would be sensible if we were to combine our efforts. We were greatly indebted to him for that. We both felt that left to ourselves we could have found the lone task too daunting for completion. The jointly authored book was published by Butterworths in January 1984 but, very sadly, Andy Crane died at the end of that year. Thankfully, however, by then it was clear that the book in which he had expended so much effort, at a time when effort cost dear, was going to be successful. He was thus able to know the sense of achievement and satisfaction in having our approach to the subject on record and welcomed, which has to be the main reward for the authors of specialised textbooks. The second edition, which I undertook alone, was a much expanded text with considerable modification in relation to the properties and use of non-metallic materials. However, in the third edition I was joined by an ex-student of the Department, who had, in fact, attended the lecture course in earlier years, Dr J.A.G. Furness (Justin), who by that time was working for the consultants Quo-Tec Ltd. The continuing development of design engineering and the growing importance of plastics, ceramics and composite materials required additional text and rewriting in many chapters. Also, since the second edition, there had been a marked growth in the availability of databases and computerised material selectors, clearly calling for a younger 'with it' contributor to the authorship! This third edition, published by Butterworth Heinemann, has recently undergone an improved reprint (1999).

Whilst my involvement with the subject was to result in the first substantial student textbook, Mike Ashby took the subject forward in his own inimitable way with the establishment of 'selection maps' and then with Cebon followed the rapid onset of computer modelling to the establishment of databases and the Cambridge Material Selector software. This was initially a teaching aid but eventually became commercially available where it is now joined to the Cambridge Process Selector, a sister programme for the identification of the way in which a component should be manufactured.

G.C. SMITH

Gerry Smith, the only member of the academic staff obstinately and modestly to remain Mr, has been a good friend throughout all my Cambridge years. He went straight from graduation into teaching, onto the staff and subsequently provided a continuity of first-rate instruction in physical metallurgy over many years. He recalls with humour, how, during World War II, a lecturer would come to Cambridge from London once a week to give two lectures to the Part II class of which Gerry was a member. The numbers were small and the students would be gathered around a table in what was then the Department Library. The said lecturer would go to the book shelves, take down 'Metals' by Carpenter and Robertson, and solemnly enquire which page they had reached on the previous visit and then essentially read and copy graphs etc. from that great, but already dated, work. No doubt the gentleman was under considerable pressure from the wartime conditions which restricted preparation and certainly Gerry learnt a great deal, although mostly, one conjectures, by concentrated endeavours outside the lectures! It may also have had a part in his development into one of the most effective teachers that the Department has had, knowing how not to do it! By nature he is quiet and somewhat retiring, although certainly not aloof or anti-social. Professionally, he has not received the recognition that he has deserved. A mark of his calibre is not only the excellence of the research carried out under his direction over many years, but the demand that there has been for his services as a referee for papers and as an examiner for theses.

Painstakingly thorough and accurate, few errors or misconceptions would pass his notice. In recent years his deep understanding of the subject and his thoroughness have served the Institute of Materials and its predecessors well, particularly in his role as Consulting Editor of *Materials Science and Technology*, most recently in harness with John Martin of the Oxford Department. Gerry was also much involved with life at Pembroke College, holding the offices of Tutor, Senior Tutor and President over the years.

After his retirement he joined me in supervision of some of my research group, to my pleasure and their advantage, and now, both as 'department geriatrics' we share the small office once occupied by our secretary. The harmony with which we have been able to work together with, I believe, complimentary strengths, has been particularly evident in the joint consultative roles that we have played, e.g. acting for civil engineers in a case concerning the fracture of fuel pipeline bends shown to be martensitic, advising in relation to the hydrogen-assisted cracking of tie bars on a bridge and the effect of edge finish on the fatigue of ribbon actuators in hi-fi speakers, to name but a few.

One of our most unusual joint roles was in providing, for a few years, a short course for the technical service staff of manufacturers of agricultural equipment, whose large light green combine harvesters are frequently to be seen on our fields. The lectures we gave were concerned particularly with the basis of mechanical properties and modes of failure, including corrosion. Afternoons were taken up with the examination of failed components brought in by the class for discussion. As could be expected, most of these failed components showed clear evidence of fatigue and the importance of design in relation to stress concentration could sometimes be demonstrated. It was somewhat reassuring that even with the generally outstanding engineering quality of the German products, occasional problems were still encountered at that time in the context of the very harsh service conditions. Maybe our activity contributed to feedback and an even better product!

Particular pleasures that we shared as the senior members of the Department were the organisation of scientific meetings and social events marking the 70th birthdays of first Alan Cottrell and then Robert Honeycombe. The IoM publications '*Advances in Physical Metallurgy*' and '*Further Developments of Metals and Ceramics*' resulted.

EXTERNAL EXAMINATIONS

Over the academic years, I acted as the External Examiner for ten different undergraduate courses at Imperial College, Sheffield Hallam, University of Strathclyde, University of Leeds and University of Sheffield, for some for more than one three-year period, all in addition, of course, to my examining duties in Cambridge. Add to that six MSc courses, Production, Processing and Fabrication (Strathclyde), Metallurgical Process Management (Sheffield Hallam), Materials Selection (London Guildhall), Archaeometallurgy (Institute of Archaeology, UCL) and Materials Science Techniques (Imperial College), and it comprised a fair amount of experience. As part of these exercises, there was usually a requirement to give a view as to the standard of the course and the response of the students in the examinations in comparison to similar courses at other universities. From written papers, this can be quite difficult to do. One cannot easily know to what extent the examination papers reflect lecture content, hand-out material or the form of questions in classes. In some quarters the value of interviews and oral examinations is denied, but I believe that, properly structured, the oral is an essential part of making both a detailed and overall assessment of the results and should always be used by external examiners. My own approach has been to table a specimen and to develop a discussion from it with individual students. If the examination were to be in mineral engineering, as at Leeds University, it would be a mineral, leading to identification, means of analysis in the field or laboratory, crystallography, sources and economic significance, comminution, concentration, smelting, application of product etc. The specimen has, of course, to be changed between students, since those having been seen would communicate with those coming later. I recall a student from Botswana, excellent in his theory papers, being nonplussed and not a little distressed when confronted by a very beautiful chalcopyrite specimen when his colleague had said that he would be asked to talk about galena and lead production. Fortunately, he used his own knowledge rather than their prompting, after initial hesitation!

For materials science, materials engineering and metallurgy students the approach would be to table components displaying a range of metals and nonmetals, requiring a deduced basis for selection, aspects of integration of selection with design, properties and their measurement, heat treatments, surface coatings etc. and branching off into likely microstructures and relationships to properties. Again, changing the components around was necessary during the oral sessions. As an old fashioned metallurgist, a particular annoyance would be when a student would not be able to identify a broken piece of enamelled cast iron, displaying both a typical grey iron fracture and enamel layers, preferring to produce it by a complicated metal and ceramic powder route as a 'composite'! When a good student is faced with such questions as these, the response is usually very positive and develops thoughtfully and naturally, reflecting proper understanding over breadth.

It has been my experience that the most significant variation between courses has been more the level of attainment at entry, as reflected in A-levels. For some, many of the students will not have exceeded grade C at A-level. This leads to an inevitable 'tail' of poor results in the degree course, which has to be dealt with year by year and may even produce some failures. On the other hand, the best students from these courses will compare favourably with any other university and there is often a clear indication of 'added value'. This seems to be particularly true with sandwich students, where greater maturity and a motivation and dedication to succeed can build on possibly less promising initial qualification.

There has also been the privilege of a great deal of experience in assessing bright school sixth formers as possible future members of St John's College to read Natural Sciences, in my office as Natural Sciences Director of Studies. Here, asking questions to reveal the ability to think fundamentally about the science of a situation requires great care and a sensitivity to the response, so that the question can be changed in its course if it is clear that no teaching has been received on a particular front. Basing discussion on an artefact in relation to physics and chemistry can again be useful.

Archaeology and Fine Art Studies

ARCHAEOMETALLURGY RESEARCH

In 1965, an event occurred which was to have a considerable impact on my metallurgical activities. Dining in Hall at St. John's one evening I sat next to a young Title 'A' Research Fellow in Archaeology, C. Renfrew (now Professor Lord Renfrew of Kaimsthorn and Disney Professor of Archaeology at Cambridge), who had taken Part I Natural Sciences and Part II Archaeology as an undergraduate. He was keenly interested in the application of front-line science to archaeological problems, with an active interest in the work that J.R. Cann, another St John's Research Fellow, had done on the analysis of obsidian glasses in the Mediterranean region. Trade routes were established for these unique materials when used as tools, each source having a unique analysis. During his Fellowship, his own PhD research, concerned with the Neolithic and Early Bronze Age cultures of the Cyclades and their external relations, was being expanded to include the independent origin and development of copper metallurgy in the Balkans. With characteristic enthusiasm, he sought my views on the likelihood of independent discovery and development and on the significance of copper–arsenic alloys, which, in that region and elsewhere, appeared between the use of copper and the development of tin bronze. Were they purposely made by the selection of specific materials or were they purely adventitious? What would the properties of the alloys have been that could have encouraged purposeful production? How like tin bronze were they? At his suggestion I supplied two addenda to papers by him, addressing some of these issues, and my interest in the long history of metallurgy was sharpened.

With attendance at Archaeometry Conferences, usually at Oxford and organised by the Research Laboratory for Archaeology and Fine Art, I made the acquaintance of many interesting people including Dr (later Professor) E.T. Hall, the Director of the Laboratory (a lively, entertaining and robust character). One of the conferences, in March 1977, was held at the Smithsonian Museum in Washington in the company of the late Professor C.S. Smith (then the doyen of archaeometallurgy), Professor Heather Lechtmann, Dr Martha Goodway, Professor Robert Maddin, Professor Jim Muhly and our own Dr R.F. Tylecote (Ronnie). Whilst at the Smithsonian meeting, I was given hospitality by Dr Fred Fraikor, who has kept in touch over the intervening years and who so kindly entertained my wife and I during a more recent stay in Golden, Colorado. In latter years, Fred has been associated with the Colorado School of Mines in various capacities. For the second phase of the 1977 conference in Philadelphia, I stayed with Bob and Odell Maddin. Bob became very active in the archaeometallurgy field, particularly after formal retirement, and we still meet from time to time. In spite of the effort already being made by metallurgists established in this field, such as C.S. Smith in the USA and Ronnie Tylecote here, I came to feel that my own particular background was well suited to the deduction of process metallurgy conditions in the manufacture of an artefact and the observed macro- and microstructure and composition of the end product (as for example reflected in the non-metallic inclusions and their relationship to the grain structure) and that a contribution could be made.

As already noted, another aspect of Colin Renfrew's interests related to the work of the St John's geologist, Johnson R. Cann, an expert on the composition and structure of magmas, including obsidian glass, characterised by a unique analysis in relation to the volcanic source. There were questions at the time, which continue to be expressed, as to the possibility that metals could also be uniquely characterised in relation to their mineral origins. Over the intervening years, archaeologists have had to be reminded of the pit-falls in relying on minor element or trace analysis other than isotopic ratios to typify metal objects in relation to one another and to a source. Being crystalline rather than a glass, like obsidian, metallic alloys frequently display substantial compositional heterogeneity. Macrosegregation during solidification can mean that one end of a casting can have a different composition to the other, a difficulty in the production of alloy spectrographic standards that had to be overcome at J. Stone (p. 44). Even the mineral sources from which the metals come can show variation in trace element analysis within one deposit, a variation which may be taken through the smelting stage and even exacerbated by variations in results from one small smelt to another.

When working with Renfrew on Balkan copper axes, it was possible to interpret macro- and microstructures to prove that these 4000BC copper artefacts were cast as a simple shape, often in an open mould, as indicated by the presence of graded concentrations of copper/copper oxide eutectic, with a core for the haft hole, followed by forging to the final complex shape. Commenting on the two papers with Renfrew, one concerning copper–arsenic alloys and the other the Balkan axes, the American archaeologist Professor Jim Muhly wrote: "these papers provide a wholly new dimension to the study of early metallurgy, identifying copper–arsenic alloy usage to be a separate part of the Bronze Age, and showing it is possible to do something with a metal object besides measuring and classifying according to type". This was a considerable encouragement to continue with work in the field, if only as a secondary activity.

A different opportunity was also made possible by Colin Renfrew when he drew my attention to a Minoan dagger from Gournia, Crete, questioning the nature of the silvery caps applied to each end of the copper rivets through the bronze tang. Already familiar with Sheffield plate, where the method of

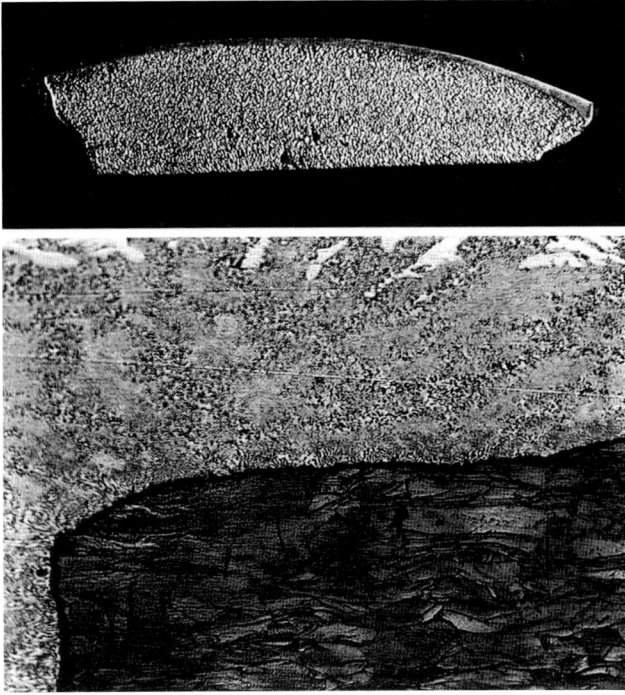

Fig. 24 Heavily etched rivet head macrosection showing final form of the silver and silver–copper cap on the rivet. The microsection shows the silver–copper eutectic at the silver/copper interface, squeezed over edge of rivet during bonding (×270).

manufacture was to hot roll copper plate sandwiched between thin silver sheet, relying on diffusion bonding at the interface (often with the generation of the silver–copper eutectic), the fact that the rivet caps were still firmly attached and protected from corrosion by the less noble copper beneath suggested a metallic bond, probably produced by diffusion in the same way. With permission to section a rivet with an attached silver cap, the macro- and microstructures showed clearly the formation of silver-copper eutectic by diffusion between the two, just as in Sheffield plate, and the way in which the final heading of the capped rivet subsequently had been achieved (Fig. 24). Thus the origin of the technique was not with Bolsover in Sheffield in 1743, but, at the latest, with the Minoans in about 1500BC. In terms of design, it was clearly evident that the outside of the rivet head had been chamfered to accommodate any squeezed-out eutectic and ensure a good capture of the silver cap.

My activities in archaeometallurgy could have expanded to fill my whole time and energy, as they had eventually for Ronnie Tylecote, but I was cautious of this happening, preferring to maintain an emphasis on research and consulting on current industrial problems. None the less, my interest in archaeometallurgy was obvious in student supervisions and over the years attracted a few research students. Thus, where a genuine proactive interest existed in any student I was always willing to give research supervision, although financial support was sometimes difficult and there are limited employment opportunities if a longer term future in archaeometallurgy is envisaged. Another problem has been that, although the work may lead to interesting new understanding in metallurgy, a context of archaeological science journals can mean that no professional metallurgists are aware of it and yet materials science journals may reject it as not being relevant in view of its archaeological content, even though this may be small.

The first of these research students in archaeometallurgy was Miss E.A. Slater (Liz), a member of one of the groups of New Hall students that I supervised in their final undergraduate year (1967–68). Her thesis, 'Metallurgical Aspects of Bronze Age Technology' was wide-ranging, to include aspects of the copper–arsenic alloy story and, importantly, practical evidence from model axe castings that bismuth segregated to such an extent in bronze that a compositional classification system for artefacts proposed by S. Jughans, E. Sangmeister and M. Schröder was untenable. I had already been aware of bismuth segregation in wrought bronzes and was aware the harm it could do from experience at J. Stone. Liz Slater's external examiner was Ronnie Tylecote, then still at the University of Newcastle. She has gone on as an archaeologist with a natural science background and is now Professor and Head of the Department of Archaeology, at Liverpool University.

The second research student in this area, in 1972, was J.A. Todd (Judith), married then to an anthropologist working in Ethiopia, and her research, divided between field and laboratory studies, concerned the Iron Age technology still being employed by the smiths of the Dimi tribe in that country. Not only was the technology for producing iron carefully observed and recorded, but non-metallic inclusion studies at Cambridge enabled associations with raw materials to be established. After her PhD, she moved back into conventional metallurgy research and is now an established Professor in the USA.

At one point Ekerhard Volker, a mature Canadian student from the University of Waterloo, Canada, joined me for almost a year, studying various aspects of lost-wax casting technology and the effect of lead on the fluidity of bronze, a topic which continues to intrigue me. Although not finally proven, it seems likely that liquid lead, produced by rejection from the initially chilled skin of bronze in which it is insoluble, greatly reduces the thermal conductivity and the suitability of the interface for nucleation from the still-flowing metal. The effect of progressively increasing lead content can be very dramatically demonstrated by a series of fluidity spirals.

Many years were to pass before my next archaeometallurgy student. In 1989, Karen Wiemer, also Canadian, started a detailed study of Roman, Viking, Anglo-

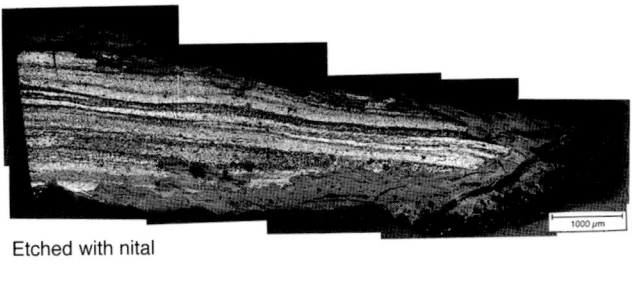

Etched with nital

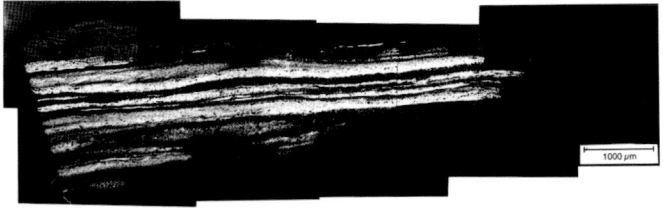

Etched with Oberhoffer's reagent

*Fig. 25 Section of Anglo-Saxon knife blade from a grave at Empingham,Rutland, UK,
showing a banded structure.*

Saxon and Mediaeval knife blades, comparing the products and the variations in technology thus revealed. On finishing, she too decided to follow a more conventional career as a physical metallurgist, initially at The Welding Institute, Abington.

Relatively close behind her, in 1993, another excellent Part II student, J. Stewart (Johnny) joined me. He made a detailed study of the iron–phosphorus system, an understanding of which was clearly important in the development of the banded structures observed by Karen Wiemer in the knife blades and revealed by Oberhoffer's etch (Fig. 25) and by probe analysis. The particular significance of the segregation of phosphorus as a result of the γ to α iron transition, not usually recognised, was established. Johnny moved to the Science Museum in London, helping, amongst other things, to promote interest and understanding in metals. All the work of my 'archaeometallurgy students' has been of a high standard and definitive in its own area, a matter of great satisfaction to me.

OTHER ARCHAEOLOGICAL INVOLVEMENT

The Separation of Seeds from Soil

For a few years, I was able to introduce a small amount of mineral dressing into the Cambridge metallurgy course, with associated practicals, for which we purchased a Wests Fagergren laboratory flotation cell and a small shaking table, especially designed and made for us by H.N. Blyth of RSM. Eventually, this material was crowded out of the syllabus as the emphasis switched more completely to Materials Science.

Imagine my delight when I was approached by two archaeologists, A.J. Legge (Tony) and H.N. Jarman (Heather), for an improved technique to enable the separation of carbonised seed from soil, rather more quickly and efficiently than wet sieving by hand. With known quantities of purposefully carbonised seeds, such as carrot and water chestnut, we were able to show a full recovery from a mixture with sand by froth flotation in the Fagergren cell, using pine oil as collector and frother. Tony Legge subsequently produced a number of field units, with sintered metal air bubblers set in a pipe grid in the bottom of a cylindrical plastic tank, which gave excellent results, and the technique is still used widely today. There is even the possible disadvantage that the quantities of seeds recovered from, say, a domestic hearth, become awkwardly large, but do at least provide a good base for subsequent sampling and more reliable results than merely recording a much smaller hand-collected quantity, where shape and size could skew results.

Determinative Mineralogy and the Origin of Metals Usage

A particular interest over many years has been the generation and honing of a theory concerning the origins of metallurgy and the sequential use of copper, copper–arsenic, copper–tin and iron, seeking to show how simple determinative mineralogy (not too different to the charcoal block tests learned at the RSM) could lead to the purposeful selection of particular minerals for the best product and how the move from one material to another could come from mineral association and the recognition of similarities and differences in colour, density, texture and changes on heating, particularly smell. For example, several arsenate products in the oxidised zone of a primary chalcopyrite ($CuS.FeS$) deposit, but containing copper arsenides (e.g. tennantite, enargite), are green in colour like malachite ($CuCO_3.Cu(OH)_2$) but produce a pronounced garlic odour on heating, which is diagnostic of arsenic generally, almost certainly leading to the use of arsenopyrite as the arsenic source when producing copper–arsenic alloy. The Homeric record describes Hephaestos, the smith god, as having several peculiarities that could be associated with arsenic poisoning! Once discovered through mineral association, probably with arsenopyrite and identified, the use of cassiterite (SnO_2) with copper to produce tin–bronze, which is not dissimilar in properties to copper–arsenic, would have prospered, perhaps almost in an evolutionary sense. In copper smelting practice, the use of iron oxide as a flux was quickly established, probably initially as a result of the limonitic gossan or 'iron hat' associated with the copper deposit. As furnaces became larger and temperatures higher, the activity of iron in the smelting system would have increased to give a higher iron content in the copper and even the separation of adventitious separate pieces of iron. Thus, the product from an all-gossan or other iron ore charge would have been established and the bloomery of the Iron Age was born.

THE INSTITUTE OF ARCHAEOLOGY, UNIVERSITY COLLEGE LONDON

As a result of an initiative by Professor Beno Rothenberg in 1973, money was raised (primarily from RTZ and VW) for the formation of an Institute for Archaeometallurgical Studies (IAMS), within the Institute of Archaeology. This was at about the time that Ronnie Tylecote had retired from his academic post in Newcastle, moved to the Oxford area and was soon working almost full-time at the Institute. An MSc course in archaeometallurgy was set up with lectures provided by Beno himself, Ronnie, Professor Bachmann (coming over from Germany), Professor Shaw (mining) from the RSM and with Dr Nigel Seeley also providing the administration and instruction in the use of analysis, conservation etc. My initial involvement came as an External Examiner for BSc and MSc courses. Ronnie Tylecote, I believe, by then realised that I was not seeking to challenge his leadership in the field and that I was acting out of interest on a very part-time basis, for initially I sensed that he perhaps viewed my activities as an intrusion. Later we became good friends and in his final illness he suggested that I should take over some of the responsibilities that he had carried. One of these I was able to do. As Visiting Professor I have taken over the lectures that he would have given for both BSc and MSc courses. My approach has been to deal with the mineralogical and physico-chemical bases of early smelting in the Michaelmas Term and then elementary physical metallurgy in relation to casting and working in the Easter Term, heavily supported by case studies illustrated with slides.

Nigel Seeley and I became good friends, co-operating on one occasion in court before Mr Justice Kennedy on a case concerning the authenticity, environmental history and ownership of a Nataraja, a large Siva bronze. The story goes that in the 12th century a notable in Pathur, Tamil Nadu, India, gave land and built a village temple that was equipped with nine bronze images. These were buried for safety, probably in the 14th century when Muslim armies invaded and briefly ruled the area. With its ruin, unlike elsewhere, the bronzes were not restored to the temple and remained buried until 1976 when a labourer accidentally discovered them when digging. He reburied eight of the images but sold one (a large Siva Nataraja bronze, depicting the god Siva dancing and set in the flaming circle) for 200 rupees. It was then sold on through several underground transactions until it ended up in London, by now provided with a false pedigree, and was sold to the Bumper Development Corporation of Alberta, Canada, for approximately £250,000, and would probably have gone on loan to a Canadian Art Gallery.

Submitted to the British Museum for conservation work, its significance was recognised and it was impounded by the police as likely to be stolen property. From the microstructural evidence, the bronze clearly had been buried for a long time, with advanced internal oxidation, although we did not believe that the degree of analytical similarity between the various bronzes in the group established that they were necessarily related in terms of a single smelting source, although they might have been. There were a number of co-plaintiffs seeking the return of the bronze from the Bumper Corporation, these being the Union of India, the local state, a public official of the temple and Siva Lingam (a cylindrical piece of stone representing a Hindu god). The inclusion of Siva himself as a plaintiff was allowable since the established medieval Hindu law, adapted by British Colonial administrators, had established that the god of the temple may act as a 'juristic person' on behalf of the religious institution, that it is, in effect, his palace and constitutes his property, which none other could own. In the end, the court found in favour of the ruined temple in Tamil Nadu. The fear then was, of course, that many religious objects from India and Asia that rest in museums might attract similar actions.

Nigel Seeley left the Institute to join the National Trust, where he now directs the conservation service. As already noted, we still co-operate from time to time, for example, in the matter of the underside corrosion of roofing lead, where our lead research at Cambridge had relevance, particularly as regards copper content and its dispersion.

Dr John Merkel was appointed to the Institute in 1988 and took on the responsibility for the archaeometallurgical courses and the in-house activities of IAMS, such as publications, and we have worked together for a number of years. There has been considerable interaction with the research students working at the Institute. One, V. Kassianidou ('Lina'), tabled some very puzzling specimens from an excavation at Huevla (Rio Tinto, Spain) associated with the working of jarosite for silver, for which only a deduced novel smelting system could provide explanation. She subsequently returned home to Cyprus, where she is now a lecturer in archaeology at the University of Nicosia, and we have kept in touch. Through her guidance I was able to see a 'gossan' or 'iron hat' almost intact above a copper working at a site in Cyprus – rare indeed. In a typical visit to the Institute, I can be called upon to comment on microspecimens relating to Islamic brass, bronze and copper objects, Japanese swords, ironmaking slags, metallic prills in copper smelting slags etc., with never a dull moment!

THE FITZWILLIAM MUSEUM

Over the years (1960–85), there was frequent co-operation with the Fitzwilliam Museum, initially with the then Keepers of Antiquities and Coins and Medals, Richard Nichols and Graham Pollard respectively, in relation to the analysis of artefacts such as Etruscan mirrors, corrosion problems, the identification of fakes etc. This activity was recognised in 1985 by an invitation from the University Council of Senate to serve as a member of the Museum Syndicate, which I did for 10 years, continuing to give metallurgical assistance when needed, but learning a great deal about the wider world of art and museum management. All good things have to come to an end however, and on reaching the age of 70 in 1996 my time as a Syndic had to finish. Three Directors had been in post during those 10 years, the late Michael Jaffeé, Simon Jervis and the present holder, Duncan Robinson, and I was delighted when the latter on behalf of the Syndicate asked me to continue to advise the Museum as Honorary Keeper of Metalwork. Such co-operation as I have been able to give has been greatly facilitated by excellent relations with the Museum staff. Whilst a Syndic, the Museum appointed a Conservator, Miss Julie Dawson (Mrs Hartley), in the Department of Antiquities and she has shown great initiative and intelligent understanding of the conservation problems as they have been encountered. Jo Dillon now gives a similar service to the Department of Fine Arts.

Work with the Department of Antiquities continued under successive Keepers, Richard Nichols, Janine Bourriau and now with Dr Eleni Vassilika, and an interest in Egyptian bronze castings developed. Interest in the detail of the manufacture of a classical bronze by the Keeper of Fine Arts and Deputy Director, Mr Robin Crighton, resulted in a γ radiation experiment being carried out at The Welding Institute, showing the armatures and successive stages of the investment casting very clearly. There is, of course, a small but excellent collection of armour and weapons also under Robin's care, so that my activity need not be wholly non-ferrous, as it tends to be in relation to antiquities. John Williams, probably the most widely experienced UK investigator of armour and its metallurgy, spent a sabbatical term in the Department of Materials Science and Metallurgy, away from his schoolmaster post, to include some studies of Fitzwilliam material, and no doubt fresh opportunities will arise in the future.

For those who are not familiar with the collections at the Fitzwilliam, which recently celebrated 150 years of free opening, it is very well worth a visit. The range of paintings and objects is extremely wide, all of the highest quality and yet one is not overwhelmed by scale. My own favourite gallery houses superb ceramics.

Consultancies

I have been extremely fortunate during the academic phase of my career in being retained by three companies over most of the period. I made it a rule that I would not undertake consultancies where there was any likelihood of a clash of loyalties. Thus my association with Timsons of Kettering has been concerned with metal failure and the selection of materials, Imperial Smelting Processes Ltd. with non-ferrous extraction metallurgy and Tube Investments Ltd., with iron and steel making and general physical metallurgy. There have been many other, shorter, associations, but only these three will be dealt with here.

TIMSONS LTD., KETTERING

I was first introduced to this company by Donald Welbourn, then Director of the Cambridge Wolfson Industrial Unit, and became a retained consultant in about 1970, a happy association which continues. The firm specialises in the manufacture of high quality printing presses, in use all over the world, but had its origins in the manufacture of boot and shoemaking machinery at the end of the last century. In 1970, Ernest, son of the founder Arthur Timson, was still very much in command as Managing Director, but later handing the role to his son-in-law Peter Brown. In spite of his age (born in 1902), Ernest was always interested in new material and material processing developments, avidly reading many technical journals and writing to authors for further explanations and comments on the applicability of their expertise to the production of components for his presses. Gear cutting, surface hardening and reduction of noise were particularly important topics always prominent in his mind. His interest in surface hardening (for gears and bearer surfaces particularly) even led us to arrange to visit Harwell to discuss the, then, new process of ion implantation. Printing presses are incredibly noisy over a wide range of frequency and separate consultation was arranged with Professor J.E. Ffowcs Williams, active in the suppression of noise, e.g. from aircraft. The only contribution I could make in relation to noise was to suggest the introduction of copper–manganese–aluminium damping alloys as 'liners' on bearing housings, but such an approach would, of course, have considerably increased costs.

In working with Ernest Timson and his staff, one felt that every aspect of each individual major machine was a subject of question and possible development, with the continuous feedback of information from machines in service. A family firm, Timsons had been built up on strong foundations, where the family, living quite simply, as prominent members of the Baptist Fuller Chapel in Kettering, had made only modest claims on profits and had always ploughed money back into the business, never borrowing or failing to make a profit, even in the difficult years. Ernest's domination of the firm, however, meant that a patriarchal style survived from a bygone age until his death in 1987. My wife and I were always very kindly invited to the Christmas dinner/dance party, where the formalities of Directors and guests sitting at a high table, processing in to take seats to a standing assembly, were still retained. He was of his age and a great man. Other Directors that I came to know were the late Charles Dickinson, John Harris, John Idell, John Jefferis, Alan Eldred, Richard Howlett and of course, Peter Brown.

My role has usually been to solve problems of material selection, perhaps where a normally specified steel is not available or where improved performance has become necessary through more demanding service or to identify causes of failure of components in service. Relatively small specialist companies, such as Timsons, have an on-going problem in relation to material supply. Trying to move outside the use of a limited range of standard steels imposes great difficulties of supply, since they are not held by stockists and direct purchase from manufacturers could only be in substantial quantity, when the requirement might only be for a few bars to produce small items such as palls. Component failure in any machinery can, of course, result from material deficiency, but is commonly associated with fatigue, where the importance of design, with the elimination of stress raisers, has often been the solution. Throughout the presses, there are many hardened components, and the prevention of cracking, in such items as gears and cylinder bearers, is an important area. It has been a happy relationship.

IMPERIAL SMELTING PROCESSES LTD., AVONMOUTH

In 1965, a 20 year association was begun with ISC/ISP as a regular consultant, attending at Avonmouth three times a year, usually for two to three days on each occasion. It began with an approach from P.M.J. Gray (Phil), the Metallurgical Research Manager, who had been a friend in my year at the RSM, inquiring on the possible merits of oxygen usage in the (zinc) blast furnace (ISF), the important new zinc smelting process, which was to become very important worldwide. He had recognised my general involvement with the topic in the iron and steel industry during my time with BOC. Following graduation, he had worked at Harwell and then in Australia on the pressure leaching of uranium ores. I had come across his work when carrying out a survey for BOC on the possible use of oxygen in pressure leaching. Over the 20 years, I came to know, as friends, so many of the staff at Avonmouth and Bristol and looked forward to my visits from both technical and social viewpoints. Since their positions and responsibilities varied over those years I will often refrain from doing other than mentioning them by name, without positions. Phil, mentioned above, is still active as an independent consultant and we meet from time to time.

In the early days of my visits, the Research Department was part of the Imperial Smelting Company (previously National Smelting Company). Imperial Smelting Processes were then separate, with the remit of selling the Imperial Smelting Furnace (ISF) technology. S.W.K. Morgan (Ken), whom I saw from time to time and who gave a great deal of impetus to the ISF process, was Research Manager and then Managing Director of ISC, to be followed in the former post by Dr J. Barbour (Joe), with Phil Gray acting under him as Metallurgical Research Manager. When Phil left ISC, the post was taken by Dr A.W. Richards (Alan).

Alan became a family friend and when Valerie accompanied me on trips to Avonmouth, she and his wife, Margaret, would spend time together. He was a very conscientious and able Bristolian and during his spell as Manager a great deal of excellent research was carried out. Sadly, he was to develop mycenia gravis, which affected both his speech and sight, and although he tried to struggle on, it was eventually to force early retirement; he did not live long afterwards.

Over the years there have been many reorganisations at Avonmouth. At one point, Imperial Smelting Processes took over the metallurgical research from ISC and the chemical research and direct links to the Avonmouth ISF were moved elsewhere. The Managing Director of ISP at this time was Dr D.A. Temple (Derek), a RSM graduate and Cambridge postgraduate, who had worked with U.R. Evans for his PhD. He usually requested a meeting and a view as to how I saw things going in the Research Department overall. As he lives in Saffron Walden, and occasionally visits the Department, we still keep in touch. What remains of ISP is no longer directly owned by RT but by International Mining Consultants Ltd and is based, with Roger Lee as Managing Director, in the Coal Research Establishment Group laboratory at Stoke Orchard. RT does not own any ISF's now; that at Avonmouth is owned by Mount Isa Mines and called Britannia Zinc Ltd.

When Joe Barbour was Research Manager of ISC there was still a section working on the development of zinc alloys and their use, involving, their physical metallurgy. The staff in this area were C.J. Swanson, J.A. Southern and Dr Chris Cross. When the physical metallurgy effort at Avonmouth was abandoned in 1972, my further contact was only with Chris, essentially a physical metallurgist by training, who had been investigating superplastic eutectic zinc–aluminium alloy which could be produced on a Hazelett continuous casting machine. He was moved to look after laboratory work on coke quality, briquetting etc., where structural studies were an important aspect of research and control. In relation to briquetting, he was associated with the chemical engineer Frank Benatt, who subsequently left to become a full-time potter! I still have contact with Chris from time to time. He continues to be based in Bristol as External Research Manager with RT Technology, responsible, amongst other things, for university contacts and contracts. Joe Barbour, himself, on moving from ISC to RTZ

Services, and joining Maurice Cahalan, became the RTZ expert and advisor on pollution control and other environmental issues.

The late L.J. Derham (Jack), often credited with the original idea and impetus behind the zinc blast furnace process, was a man who was never happier than when directly involved in experiments and trials, determinedly seeking practical solutions to unsolved problems in process development. Our association was largely concerned with 'brain storming sessions' when he would bounce one idea after another off the 'academic from Cambridge', to, I hope, our mutual respect. J. Lumsden, FREng, was the resident chemical thermodynamics expert, much revered by the rest of the staff and of international reputation. His conclusions as to the viability of a process were always significant in relation to contemplated development, but he could be very difficult to talk to. To some extent this was also true of S.E. Woods, FRS, whom I knew over a much longer period of time. My sessions with him were mainly listening, trying to interject comment as he rehearsed fundamental thermodynamics and thought in relation to processes, particularly the operation of the zinc blast furnace. As regards oxygen injection, I recall that it was our joint conclusion that if the furnace condenser gas could be treated to recover CO, to be fed back to the tuyeres with oxygen, there could be substantial savings in coke requirements per unit of zinc, together with a significant increase in the output from a given sized furnace.

As well as Phil Gray, Dick Healey, Alan Self and Colin Harris had all also been members of the 1947 RSM metallurgy class. Dick, I recall, was involved at one time with blast furnace calculations and modelling, before he retired early and went into the antiques business. Alan Self I saw only occasionally in a social context since he was not in the Research Department, but worked on the plant. Colin Harris was a prime contact throughout the 20 year period. After military service, he joined Imperial Smelting in about 1949, in the early days of the then new zinc blast furnace process and built up a wealth of experience and understanding of the ISF, and on the retirement through the ill health of A.W. Richards, became the Department Manager. He is a direct, robust character and we 'got on' well.

W. Hopkin (Bill), an astute Scot, had established a position as the hydrometallurgy/electrometallurgy expert in the team and dealt with such matters as copper dross leaching, for which a pilot plant was established. It was natural, also, that there would be chemical engineering activities and these were normally undertaken by M.W. Gammon (Mike) and P.J. Gabb (Phil). Both were excellent men. Frequently, the strength of the team would be called to help with plant start-up duty where an ISP-designed process was being installed elsewhere. After my time at Avonmouth had ended, Phil Gabb spent a number of years at Kennecott in the USA, although not in relation to ISP designed plant. Mike Gammon remains in Avonmouth with the surviving ISF.

In the early days, a member of ISC staff, with whom I had discussions, was Dr D.J. Fray (Derek), a graduate from the RSM. He was, and is, a first class man and when he applied for a lectureship in the Cambridge Department, I was happy to add my whole-hearted support for his selection. Thus, he became a colleague, but moved for a few years to Leeds University as Professor and Head of the Department of Mineral Engineering, returning to Cambridge as Professor and a colleague again. There was much truth in the view that the best way to a Cambridge chair was to move away and then return!

So much for the people that I recall most clearly. A visit would have a prepared programme of discussions with different people over the three days, with topics relating to the work in hand at Avonmouth, both in the research laboratory and that taking place on major projects, in co-operation with staff on the plant design and operation side. Examples of these were: design and operation of pilot plant copper drossing furnace for ISF Bullion; continuous lead decopperising to low level with sulphur; continuous lead refining for removal of arsenic, antimony and silver in a cascade of stirred tanks; vacuum pilot plant design and operation to investigate lead flow and zinc distillation; redesign of vacuum dezincing (VDZ) semi-commercial plant on No. 2 ISF; and the design, installation and operation of the first commercial VDZ plant on the Swansea ISF. On the Research Department side, the VDZ process was strongly supported by N.A. Warner (Noel), an Australian chemical engineer/metallurgist who later became Professor of Mineral Engineering at Birmingham University. Other ISP staff involved with these major projects included A. Robson (Alf), John Castle, Dr J.R.A. Davey, Dr Harvey Richards and Dr I. Matthews (Ian).

My role was always to listen and make any comment or suggestion that I could. There would also be a short description of any relevant work going on in my group at Cambridge. Looking back the comment could be made that the academic research (and maybe many of my contributions at Avonmouth) were mainly aspirations and good ideas in principle for ISP, that did not make it in the rough and tumble of commercial practice. However, the stimulation that the meetings provided seemed to have been useful on both sides and, as far as the academic research was concerned, any commercial spin off would have been a bonus added to the development of the students during their postgraduate years.

As with other consultancies, my activities at Avonmouth led to a number of research projects being supported at Cambridge. The work of Geoff Price, David Canham, Alan Begg, Ted Wright, Bob Cochrane and Stuart Sarson is described elsewhere. One of the Johnian undergraduates that I supervised, Bill Hunter, joined RTZ on graduation and, when I last visited, was involved in plant management at Avonmouth.

Assessment of Mineral Assemblies

In working with ISP, by 1985 I was of the view that developments in physical methods of compositional and structural microanalysis and image analysis could have taken us to the point mineral technologists and extractive metallurgists had been working towards, and long dreamed of, where we could visualise a rapid and largely automatic capability in the assessment of mineral assemblies, whether these were ores, concentrates, tailings, drosses or slags. In the extraction of a value from an ore body, the route that is followed through comminution, concentration and extraction to use is fundamentally dictated by the following characteristics:

(1) level of value in the ore as part of the total analysis,

(2) dispersion of that overall content between mineral phases,

(3) size ranges of the various phases present,

(4) nature of the interfaces between phases, e.g. degree of interlock,

(5) defect concentration within mineral grains affecting ease of comminution,

(6) degree of complexity of the association of the value within mineral phases which would affect the extraction route,

(7) presence of specific impurities in a phase, which would affect surface properties and thus physical separation characteristics.

If all these features could be assessed quantitatively using the latest techniques, then it is not out of the question that a rapid computerised evaluation programme could be developed to enable a given material to be assessed predictively in terms of the preferred routes for extractive treatment and whether any of the routes could be economically viable under given local conditions of, for example, labour and energy costs.

There had been many instances where a knowledge of the way in which the overall composition was divided between phases within a given material, usually assessed by electron probe microanalysis, had given important information relating to the use of a particular processing route. Examples were the identification of phases in solidified copper matte, drosses and slags cooled at varying rates for assessment in relation to both chemical and physical processing, with important results, gold dispersion in dumped materials and the effect of the dispersion of iron on leachability when upgrading a residue to TiO_2.

Interestingly, a recent publication by Stephen Handley describes the work of the Natural History Museum (NHM) consultancy services in relation to analysis of metal residence, pointing out that mineralogical insights could lead to cleaner or more efficient ore extraction technologies. Things currently thrown onto the spoil heap could be recovered if only we knew specifically where the minerals were. Describing a project for a ferronickel source in South America, they found that the nickel was in a series of layered silicate (clay) minerals, but in certain parts was "actually within oxide phases". One of the techniques employed by the

Natural History Museum was X-ray diffraction, enabling more reliable quantitative analyses of mineral mixtures and then using electron probe microanalysis to find out where the nickel was, as an oxide or as a silicate.

Electron probe microanalysis has, however, several disadvantages. It cannot give a full analysis to include all elements and increasing difficulty is encountered with elements of lower atomic number. The lower limits of detection are of the order of 100 ppm depending, of course, on the particular element, which is not sufficiently sensitive for many purposes where a precious metal value or a harmful impurity could be significantly present at lower levels than this. The instant total energy dispersive analysis, employed in most present day scanning electron microscopes, does not give high levels of accuracy, which requires X-ray quanta to be analysed in spectrometers, a rather more tedious operation.

Unfortunately, the Research Department at Avonmouth was closed down and the opportunity for me to take the matter further was lost, although developments in the area are apparently proceeding within RT.

TUBE INVESTMENTS LTD.

In the late 1950s, I had given several lectures on oxygen in steelmaking to groups within BOC and to others such as the Goldsmith's Society in the Department of Metallurgy at Cambridge. This had been arranged by the late C.N. Walters (Cedric), then a student at Cambridge, who had worked with me in the research foundry at BOC as a vacation student. Once ensconced in Cambridge, I was invited to give a talk at TI Research Laboratories, Hinxton Hall, associated with a staff dinner, which I believe was concerned with the history of steelmaking and its development into the use of oxygen. As a result, I was invited by Dr (now Sir) James Menter, Research Director, to accept the position of consultant to the Laboratories, with special reference to research and development on the production and use of steel, this at a time when Park Gate Iron and Steels, then part of the TI Group, were installing a Kaldo oxygen converter. Whilst elegant and flexible in relation to charges and refining, the process suffered the almost inevitable mechanical weaknesses introduced by the introduction of rotation as well as tilting in such heavy equipment, very easily affected by slag overflow. Increase in size over the original Swedish plant also gave greater circumferential speeds and increased refractory wear. In the end the process was not widely successful.

It was at this point that I met Dr D.A. Melford, (David) a Cambridge graduate, postgraduate and outstanding metallurgist (later to be awarded a well-deserved OBE and FREng), who was to become a good friend. As described elsewhere, an initial research area for me at Cambridge was to be the formation and behaviour of non-metallic inclusions in steel, using a Cambridge Instrument Company SEM, Microscan I. This commercial instrument owed a very great deal to the work of Dr P. Duncumb (Peter) and, particularly in its engineering design, to David Melford. Our mutual interests naturally brought us together. My role at TI was, as at ISP, to discuss research ideas and project progress with members of the staff, to act as a 'sounding board' and to help and advise in any way that I could. In one or two cases, it resulted in a member of TI staff being seconded to work in the Cambridge Department for a PhD, with Jim Tolley and Colin Honess being the most significant. In one instance my own work on the occlusivity of hydrogen in steel was to coincide with an investigation into the failure of a hydrogen cylinder produced within the then TI group, bringing together representatives from both BOC, TI and myself.

My interests and metallurgical approach were very much in sympathy with those of David Melford and when he later became chairman of the DTI Metals Minerals and Reclamation Committee I accepted his invitation (1987) to serve on the committee with alacrity. This committee was eventually subsumed into the Industrial Materials Market Division Technical Advisory Committee, of which I also became a member, serving under both David Melford and Alan Parker, Managing Director of T&N Research Centre, Cawston Hall. When the latter retired, he was succeeded at Cawston by Alan Begg, whose postgraduate work with me is described elsewhere.

Sadly, TI closed R&D at Hinxton Hall in a restructuring exercise and the companies on whose commercial activities my work had a bearing are no longer members of the TI group. Nevertheless, I still meet several members of the staff from the days of this happy association, David Melford, who lives fairly close by, Peter Duncumb, Don Pashley and Mike Stowell, all now of course 'retired' but still scientifically active.

Beyond Cambridge

FURTHER RELATIONSHIPS WITH
THE ROYAL SCHOOL OF MINES

Contact with my Alma Mater has been maintained throughout my career, initially through visiting to give an invited lecture or to see staff when in London on business and then, once an academic, in the course of professional activities, principally as External Examiner for individual MSc and PhD students, but also for the BSc (Eng) undergraduate courses in two three-year spells and recently for a period as examiner for the MSc course on Research Techniques in Materials Science. This one-year course was also attended by foreign exchange students, mainly French, for their Certificate of Study, who impressed me greatly by both the quality of their work and by the clarity of their presentations, in English.

It was in my continuing relationship with the RSM, however, that I may have made a major mistake. I had earlier been offered, but not accepted for several reasons, the Chair in Mineral Engineering at Birmingham following the retirement of Professor Stacey Ward and then, around 1980, I was approached unofficially by Professor Geoff Ball, who had succeeded Professor Dannatt, to see if I would be interested in succeeding him as Head of the Department of Metallurgy at the RSM on his impending retirement.

There had been some friction within the Department, almost on a chemical versus physical metallurgy basis, and it was apparently the general view amongst the staff that if I were to be elected the breadth of my interests and experience and my friendship with so many of them, whether on the chemical or physical side, would be a unifying move. Clandestine café meetings with Professor Ball were followed by a formal meeting with the Rector, and I was offered the post. My wife and I considered this challenging and exciting opportunity for me to lead an excellent team at my Alma Mater very seriously but, unfortunately, the salary offered was totally unattractive. At that time, West London house prices were considerably higher than Cambridge (not true today perhaps). No doubt I could have found some further consultative work, but at 54 I really needed very positive inducement to resign my Readership and College Fellowship and leave Cambridge.

Some of the staff at the RSM may have felt that I had let them down, having pressed for my appointment, and maybe all would have turned out well in the end – who can say. After a while the post was taken by Dr Don Pashley, FRS, from TI Research Laboratories, Hinxton, who I already knew well from my consultative association with the staff there.

PROFESSIONAL ACTIVITIES

I joined the Institution of Metallurgists as a student member and with accumulating years of experience and responsibility moved through the grades of membership, becoming a Fellow in 1962. In 1958, I also became a Member of the Institution of Mining and Metallurgy, the other professional body in the metallurgical area but dealing with non-ferrous extraction metallurgy, when my activities at BOC had extended into the use of industrial gases in non-ferrous smelting processes. Their journals and membership of the Iron and Steel Institute and Institute of Metals provided the learned society background which I have always believed to be vitally important in continuing education. In fact, as noted earlier, it was a requirement at J. Stone that R&D staff should read journals as circulated to them in the daily postbag.

From an early stage, however, I was conscious of the considerable degree of overlap between the functions and administration of the professional body and the learned societies, even to the extent of rising to my feet at an AGM of the Institution of Metallurgists in the late 1940s or early 1950s and expressing a view that amalgamation of the various bodies would be of great benefit to the profession, giving a bigger impact on both industry and Government, and to individual members whose metallurgical interests were often not neatly pigeon-holed and who probably needed to join more than one learned society, adding also that amalgamation could provide better London facilities than any of the individual bodies. The President of the Institution in the Chair was none other than Professor A.J. Murphy, my boss in the earliest days at J. Stone and Co. Ltd., who dismissed the whole notion as preposterous. He regarded the idea that learned society and professional functions could be combined as totally unacceptable, although such combined activities existed, in essence, in the functions of the Institution of Mining and Metallurgy for those in the non-ferrous extraction field. The unification of the professional and learned society institutions serving metallurgists was also recommended in my report to the Science Research Council in 1969, as noted later.

It was thus that when, many years later, the Iron and Steel Institute and the Institute of Metals joined to form the Metals Society, with headquarters at 1 Carlton House Terrace, I was delighted, although the Institution of Metallurgists was still separate. The Metals Society activities were divided into metal science, metal technology and iron- and steelmaking. The equivalent extraction aspects to iron- and steelmaking on the non-ferrous side were still covered by the IMM.

I joined the Metals Technology Activity Group of the Society in 1975 and continued as Chairman from 1977 until 1979, taking responsibility for conferences in the subject area. With the conference 'Solidification Technology in the Foundry and Casthouse' (15-17 September 1980, Warwick Institute), I learned the hard way just how much personal effort is required to arrange and manage meetings of that size (about 90 papers) and in those days there were really quite few support staff. By 1979 I had become a Council member of the Society, and Vice President on the Executive Committee in 1982. At the end of 1977, my friend and colleague at Cambridge, John Knott, who was Chairman of Activity Group 1 (Metal Science) joined me, as Chairman of Activity Group 2 (Metal Technology), in a joint formal letter to the President of the Metal Society suggesting that there should be amalgamation of the Society and the Institution of Metallurgists, a view held by us both and by a majority of members on our panels.

The response from Professor Jack Nutting, who was President of the Metals Society at the time, was a 'dusty' one. He believed that the Joint Liaison Committee between the Institution and the Society that had been set up as a result of a policy review carried out in 1977 served our requirements adequately and said "I would hope that the way ahead in the future is not in the amalgamation of the two Societies along the lines you suggest, but in the identification of the special areas of the two Societies and an area where there can be friendly and helpful co-operation on both sides. All the signs are that this is developing and I would hope, therefore, that you would feel it wiser for me not to submit your memorandum to Council through the Executive Committee."

After careful consideration of this rebuff, John and I felt that we should not insist on the matter being pursued further at that time. However, the issue would not go away and the ground swell of opinion for change continued. After consideration by a committee chaired by Robert Barnes, which was unable to make substantial progress but which believed that, if possible, amalgamation should take place, a Working Party was formed in 1980 under the chairmanship of J.O. Hitchcock (John) who had retired from INCO, but had become Chairman of Engelhard, UK. It was a purposefully small committee on which the Metals Society was represented by Arthur Whiting, my ex-colleague from BOC days, but who was by then managing director of Davy United, Nevil Lee (Arthur Lee, Sheffield steelmakers) and myself. The Institution of Metallurgists was represented by Jim Bradbury (President 1978–79) from INCO, Roy Johnston, BNF, and Professor Jack Nutting. Sir Geoffrey Ford, then Secretary of the Metals Society, was secretary of the committee. It having been established that there was, in fact, a desire to amalgamate, Jack had joined forces! The committee quickly decided that amalgamation was a definite commitment and that, whatever difficulties arose, these had to be overcome, however long it took. In fact, negotiations lasted from 1980–83, a year longer than originally anticipated, with many meetings.

Thus, my dreams as a youth some thirty years earlier of one organisation serving the metallurgical profession had become reality and I felt honoured to have had some part in it, although I had not worked to that end for some time until the joint letter with John Knott to Jack Nutting, which so shortly preceded a swing in attitude by the hierarchy of the Metals Society. Following the formal amalgamation in 1985 to give the new Institute of Metals, I continued as a member of the Executive Committee and Vice President until 1987. The Institution of Mining and Metallurgy remained isolated and I tried to bring about consideration of them joining the new Institute when still on the Executive Committee with Sir Hugh Ford as President. Some contacts were made but nothing came of it, largely because their membership was mainly overseas. In the present this might not now be seen as a barrier, although the IMM seems to have been moving ever closer to the mining interests.

As Chairman of the Editorial Panel for *Metals Technology* from 1977–84, I was also asked to chair meetings of a journal spectrum sub-committee for the new Institute of Metals. Consideration was given to combining journals into a 'Proceedings', with subdivision and purchase according to members' interest, much as with the Institution of Mining and Metallurgy. Although supported by the sub-committee this approach did not eventually find favour with most of the existing journal committees, nor was it favoured by the marketing staff. I still believe that such an approach could have attractions for the Institute of Materials.

In the new spectrum, my own interests were transferred from *Metals Technology* to *Metal Science and Technology*, but giving way to David West as Chairman of the Editorial Panel. I have, however, continued participation as an Associate Chairman since 1985, a role shared with old friends, Professor Ray Smallman and Professor Mike Stowell. With the institutional amalgamations that followed to form the Institute of Materials, *Metal Science and Technology* simply changed its name to *Materials Science and Technology* with its remit and Editorial Panel membership suitably widened. David West has, in recent years and particularly since retirement, taken a large and very valuable part in Institute of Materials affairs. My own professional involvement with the Institutes before I 'fell off the perch' was largely in the 1970s and 1980s, when journeys to London were probably more frequent than academically desirable.

In the Metals Society days, I recall one particularly pleasant event when my wife and I joined the mission to Scandinavia, which visited Finland, Sweden and Norway. The mission was led by the then President, Michael Dowding of Davy United and included John Hitchcock, Geoffrey T. Harris, Bob Hewitt, Nick Younger and the steelworks consultant, Alan Pengelly. Bob Wood, Secretary of the Society, carried all the administration excellently, as he did at 1 Carlton House Terrace also. It was Alan's practice to take his swimming trunks with him wherever he went, usually with swimming restricted to hotel pools, where his entry into the water was designed to soak onlookers! In Norway, however, he took to swimming in the fjords and in Christiansand persuaded me to join him. The inward dive from rocks was a severe shock, I had never swum in such cold water,

*Fig. 26 At Elkem-Spigerverket, Norway with John Einerkjaer, Geoff Harris
and Alan Pengelly.*

and getting back out onto the rocks was difficult. After this I refused his invitation to such exercise after an early call in Bergen, but I am told the Norwegians coming back in their fishing boats to harbour were astounded to see this bearded Cornish gentleman bound for the open sea, in even colder water, and saluted him with their hooters! I had believed that Alan had retired to the West Country to make furniture, but from a recent copy of *Steel Forum*, I see that he is still metallurgically very active having organised and chaired a major meeting. We had a particularly good time in Norway, where a great programme had been organised by John Einerkjaer of Elkem-Spigerverket on the Norwegian side. Fig. 26 shows myself, Geoff Harris and John Einerkjaer. Alan Pengelly is in the background.

Michael Dowding was an engineer, not a metallurgist, and kept me close during the mission so that I could explain what was being said or seen, if required. When visiting me once later in St John's College for lunch he wore a loud check jacket and was mistaken for my bookmaker by other Fellows! He knew that I was involved in College investment activities at the time and maybe it was this that prompted him to invite my wife and myself to dinner at his Drayton Gardens, Kensington flat. We felt very much out of place since all the other male guests

Fig. 27 The TI Hinxton table at the 1969 annual Iron and Steel Institute Dinner. Present (amongst others) are David Melford, Charles Holden, Geoff Wynne, John Sawkill, Jim Menter and Denys Richardson.

were associated with merchant banks, as was Mike by that time. Amongst them was Campbell Adamson who had to be reprimanded by Elizabeth Dowding for knocking his pipe out into the imitation (gas) log fire! Eventual knowledge that the impressive butler had been hired for the evening was somewhat reassuring.

The annual dinners of the Iron and Steel Institute and then the Metals Society were grand affairs, and no doubt still are. As an academic it would have been difficult to consider attendance but for the kindness of companies, particularly in my case Tube Investments Research. A photograph of one of these events is shown in Fig. 27. Our table on that occasion appears to have been outside the hurly-burly of the central dining area of the Great Room at the Grosvenor House Hotel!

In recent years, the major change has been the formation of the Institute of Materials through the amalgamation of the Institute of Metals, the Plastics and Rubber Institute, the British Composites Society and the Institute of Ceramics, in which I had no part to play. Some may have seen it as an amalgamation too far. Although not enthusiastic for the development, in my view in the context of the integration of materials use and design on the ground by engineers and the way in which teaching in the subject has been integrated to a common applied physics and chemistry base, with less detailed treatment of any one material, e.g. metals, amalgamation was probably going to be inevitable. In any case, divisional activities with clear objectives and devolved responsibilities and authority within the framework of the amalgamated body can, and I believe must, continue to serve more specialised interests, as for example already established with iron- and steelmaking and most recently introduced for castings and surface engineering, a most welcome development. A particular difficulty lies in the journal field where it has been difficult to persuade those working primarily with plastics and ceramics that *Materials Science and Technology* is a suitable vehicle for their papers, usually seeming to prefer the specialisation of material-specific publications. Thus, so far, the majority of papers in *MST* continue to be concerning metals, as in the days of its forerunners. Perhaps it is time to admit the situation and rename it *Metals Science and Technology*!

ACTIVITY FOR GOVERNMENT

Looking back, I seem to have spent a considerable time supporting the SERC (SRC) and DTI, but there is no doubt that I benefited in many ways from the experience and the friends made. My first contribution was the preparation of a survey of relationships between the metallurgical industry and the Universities, prepared in 1969. Companies and universities were visited in the preparation of the survey, with particular emphasis on the quality of graduates and the adequacy of their preparation for industries' needs, the subsequent training of this manpower in industry and the type of research that industry would like to see carried out in universities. The report contains much that is now history, such as a widespread desire in both academia and industry for a genuine unification of the institutional bodies representing the profession. The next contribution was to a Materials Processing Sub-Committee of the SERC during the 1970s, followed by a four year stint (1980–84) on the Science-based Archaeology Committee, under the Chairmanship of Professor Richard West, a botanist, also of Cambridge University. As he lives in my own village, we have continued to meet on the street fairly frequently, but would not have recognised each other before. His excellent Chairmanship and the very interesting group of people brought together for the purpose, ensured that these were very enjoyable meetings, although there was, throughout, a shortage of really good proposals at that time that merited support. In the metallurgical field, there were clearly few with in-depth understanding of the science involved who were interested in pursuing research in the area and if science is to be applied to archaeology, it should not be half-baked, as regrettably some has been. With a growing acknowledgement of the importance of science in archaeology and its penetration into archaeology courses it may be that the situation has improved.

In 1987, I became a member of the DTI Metals, Minerals and Reclamation Committee, under the chairmanship of Dr David Melford of Tube Investments Research, Hinxton, of whom more is written elsewhere. This committee was eventually subsumed into another, the Industrial Materials Market Division Technical Advisory Committee, initially with David continuing as Chairman, succeeded by Dr Alan Parker, Managing Director of the then Turner and Newall Research Laboratories at Cawston, near Derby. The job of the Committee was to advise the DTI in relation to materials issues and on applications for support for research and development in the area by industrial companies. In spite of the extra workload that membership entailed, particularly in dealing with the considerable amount of paper involved and background reading, the contact with so many old and new friends from industry was enlivening and enjoyable. Obviously those members of the Committee with strongly metallurgical interests stand out in memory, such as Peter Hughes, then Managing Director of Glencast Ltd., now Chief Executive, Scottish Engineering. Peter had effected a management buy-out of an ailing steel foundry at Leven in Fife and who, by personal commitment, business acumen, energy and leadership had turned the business round, using a great deal of metallurgical know-how in the improvement of products; he has continued to be a very kind friend over the years. British Steel was ably represented by Bob Baker, then Head of the Swinden laboratories and by Mick May from Port Talbot. Nicholas Younger, an old friend through consultancy and from his Cambridge days, represented the Davy group, as did J.B. Scuffam (Barry). David Driver represented Rolls-Royce for some of that time, always bringing sound common sense as well as metallurgical ability to the meetings. On the DTI side, my main contact was Bryan Kerrison, who had been a steelmaker in earlier days, but on the Committee there were also Robert McVickers, Craig Octon, Walter Vickers, Alan Whitehouse, Tom Sinclair, Robert Higman and others, representation varying with time and in-house responsibilities.

The rules under which we worked varied somewhat with the political mood. At one time policy would be to favour ideas remote from market (i.e. long-term research) and then near to market. Attitudes would change from favouring market pull to proactive invitation for work to be carried out in a particular area. Small to medium businesses with important research needs could be favoured, but it had to be recognised that the capability for carrying out research effectively more often lay with the staff and facilities at major companies. My advice was always to identify potential winners rather than be vigorously committed to a restrictive policy of any sort. The important factors should be the quality of the people involved in presenting cases in terms of commitment and track record and the likelihood of there being a successful outcome, technically and commercially.

Clearly, the majority of the DTI staff did not have the technical qualifications for detailed research assessment, which was our role. With the arrival of a new President of the Board of Trade, the policy of supporting industrial research in this way was abandoned and the Committee was scrapped. Several of the DTI staff concerned retired at about the same time. I have since felt that there may not have been adequate liaison during the progress of grants and follow up for dissemination on the completion of a project.

One of the proposals initially presented to the Materials Advisory Committee by Gordon Higginbotham, then at Rolls-Royce but now an independent consultant (Concurrent Technologies Transfer), was aimed at generating a national computer simulation programme for the design and production of castings which could be available to foundries. It was felt that whilst some work was going on in the UK, we were falling behind the USA and others in Europe in this field, which could have a serious effect on the future of our foundry industry. It was agreed that a structured programme (CAST), of which I was to be the coordinator, should be established to improve the situation, with Phase I as a 'catching up' exercise, establishing the availability, capability and use of existing programmes in the UK, and then Phase 2 to try to bring together a number of companies with substantial existing expertise to produce a new modelling capability which could be made available nationally. The participants in Phase I included FOSECO, Ove Arup and Partners, Kent Aerospace Castings, British Cast Iron Research Association (BCIRA) and J. Stone and Co. Ltd. The co-operating companies in Phase 2 were Alcan, Rolls-Royce, Turner and Newall and Finite Element Graphics (FEGS). The latter are based in Oakington, Cambridge and hence I came to know the founder, Geoff Butlin, and the manager of their involvement with CAST, John Rawlinson, quite well.

The structure of Phase 2 was to develop a core programme based on the FAM analysis package, marketed by FEGS Ltd., and then to connect to it satellite programmes dealing with such aspects as geometry transfer, hexagonal meshing, foundryman's interface, database structure, parallel processing etc. as the need for them was recognised, these to be separately funded and involving the co-operation of others beyond those involved in the core programme. We foresaw that there could have been 10 of these, but with a change of DTI policy in 1993, when no more support for co-operative industrial research was to be forthcoming, only by protracted special pleading were we allowed to proceed with two of the satellites, VERICAST (FEGS, Rolls-Royce, T&N Technology Ltd, AETC and Ove Arup) and CAPTOOLS (Professor Mark Cross of the University of Greenwich, FEGS, PARSYS, Transtech and Caplin).

The Phase 2 core programme partners at Alcan were mainly interested in modelling the direct casting method for producing aluminium alloy billets, and were also developing their own model, which led to their withdrawal before the satellite programmes began. In Phase 1 of VERICAST, three simple shapes, a bar, T plate and simplified turbine blade, were cast under carefully controlled conditions and modelled in FAM. The casting conditions, alloy purity details and cast results for the T plate were made available widely, enabling their use for bench marking on available modelling programmes. In Phase 2 of VERICAST, the remaining industrial partners, both with a primary interest in modelling the solidification of turbine blades, worked successfully towards the verification of the programme that had been developed with FEGS. In CAPTOOLS, Mark Cross led a University of Greenwich team developing the parallisation codes for solidification modelling to give more rapid output, supported by contribution from hardware producers. Like so many others internationally, his team was also developing a computer simulation package, in his case known as PHYSICA, which was also showing considerable promise. Throughout the CAST projects, I was always conscious that to a degree each of the participants, whilst co-operating with one another as fully as required by their contracts, also had their own agendas.

Some papers for conferences and information on VERICAST were published and distributed and very substantial reports for the DTI were prepared on all the stages of the project, the latter presumably still languishing somewhere in Buckingham Palace Road. Looking back, it was sad that the project was not allowed to go further forward. A more consistent policy in relation to industrial research by Government would have ensured an even better outcome.

Over the years, there was steady contact with such Government agencies as AERE, Harwell, NGTE, Farnborough and NPL, Teddington. In the latter case, in the context of solidification modelling, it was clear that high temperature data for the physical properties of the liquids and solids involved was often not available or was unreliable, e.g. heat capacities, enthalpies, melting ranges, fluid flow density, viscosities, surface tensions, thermal conductivities in solids and liquids, physical and mechanical properties of liquid/solid mixtures. The same was true in other areas where the modelling of high temperature processing was involved, e.g. welding and forging. NPL put in place programmes to provide this type of data, developing the measurement techniques needed. As co-ordinator of the CAST programme I naturally became involved.

A good many visits were also made to NPL in relation to programmes concerning the testing of electronic assemblies, particularly in relation to soldered joints. Although this was obviously related to the work at Cambridge carried out by Nick Green, my official role was as 'back-up' to Professor David Hills, whom I knew very well from our happy association at Sheffield Hallam University in the many years that I was external examiner there. By the time of the NPL work, David had retired and was working for the DTI as a consultant on the programme.

Conclusion

INDUSTRIAL DECLINE

In the introduction of this work, there was reference to the fact that during my professional life there had been a steady decline in the UK industrial manufacturing base, of which metallurgy was so much a part, particularly heavy engineering, and I have also seen so many of our historic companies disappear or change ownership to foreign competitors. Even the great BOC has been broken up, with the pieces taken by L'Air Liquide of France and Air Products of the USA, after 100 years of UK ownership. Thus, although my own personal experiences as a metallurgist have always been enjoyable, I have despaired at the way in which much of our industrial heritage seems to have been frittered away. Whilst my views may be over simplistic and much is now obscured by globalisation and international links, it seems to me, that in the past, we have been badly let down by successive top managements, out-of-date relationships with the workforce and by governments. Whilst possessing scientific and technological excellence, business excellence was frequently not present. Accepting that overseas competition with cheap labour may have undercut some industries, there is surely no excuse for the loss of control of British firms through foreign buy-outs of failing businesses which are then revitalised. How is it that there are now no major British-owned car manufacturers? The fact that both French and German-owned plants seem to be prospering here and that the Japanese are so successful, has surely undermined any argument about labour costs. The importance of UK Ltd. seems also to count for little with some managements and investors.

As Dannatt indicated over 50 years ago in a lecture at Imperial College
"The output of industry provides additional capital or goods and it is used in various ways. Part, or its money equivalent, is paid in wages and salaries, part in raw materials and supplies and part in the maintenance of plant and charges for re-equipment, depreciation and the like. The surplus is profit, which goes to swell the total capital, or material wealth, of the community. The building of a new plant calls for considerable expenditure and use of material, for which immediate payment must be made. These payments must come from industries that are in profit; they cannot come from industries that are run at a loss. It is immaterial whether the profit is distributed as interest to individual shareholders, as additional wages or salaries in a bonus scheme for employees or whether it is taken by the government from a nationalised industry, <u>the necessary proportion of it must be made available for new ventures or otherwise industrial activity will stagnate and eventually die</u>."

My impression has been that in the past, too little effort was made in profitable times in this respect of returning profit to company updates, i.e. too little investment that did not involve heavy borrowing, putting the company in the hands of financial institutions and other investors on the stock market who care little about the business itself.

I have also been bewildered by the steady reduction in spending on research and development that has often been a primary move in trying to cut costs and go into profit, rather than aiming for growth and recognising the part that research has to play. It may well be true that research and development expenditure is difficult to balance in short term accountancy, but cutting off contact from an understanding of the scientific and technological advances being made elsewhere is extremely dangerous, where R&D acts as a 'window in the world', quite apart from losing the ability to have in-house trouble-shooting and inventive capability. Nor is it a tap that can be turned off and on at will. It takes a substantial amount of time for any research team to get 'up to speed' in a specialised area and, in any case, the national picture will not allow for a reserve of suitably qualified staff to be available. This is particularly true with the demise or down-sizing of many centralised government laboratories, research associations and large industrial research laboratories, which quite apart from their central role in research, were a training ground for so many young graduates in metallurgy and then materials science. Since the metallurgy content of degree courses in materials departments at Universities is now so much reduced, the loss of these laboratories, where metallurgical training could be completed in the context of real industrial problems, has had a serious overall effect. In the best cases, they provided a preparation for young people before going into plants to effect metallurgical management. Notable exceptions to the widespread shrinkage has been the British Steel Castings Research Association in Sheffield, now after several changes of name, the Castings Research Centre, and The Welding Institute (TWI), previously known as the British Welding Research Association. Both are now international centres of excellence with broader technological remits.

In some quarters, it is almost as though a policy was being adopted based on the notion that the aim should be not to be first but to be second and that to be first in anything was likely to be too costly. Such a policy can, in the end, only lead to subservience and decay. I do not claim any expertise in economics but it must be significant that those countries that have outstripped us in technological commerce spend a larger proportion of their gross national product on research and development. The UK 'R&D Scoreboard' for 1999 published by the DTI shows that in the engineering and machinery sector only a few companies invested in research at more than 2% of sales, with an aggregate of 1.6%, as compared with the international R&D intensity of 3.3% in this field. Star performers were in the pharmaceuticals sector where the UK R&D intensity is 15% and leads the world. The individual figure for the then British Steel was 0.7% and for British Alcan 0.6%. No doubt there are reasons for this, but it hardly seems encouraging.

Perhaps another basic fault lies in an historic inability to use research effort properly, which has led to its dismissal as being of dubious value. I recall D.J.O. Brandt (Oscar), as editor of the *Iron and Coal Trades Review*, in the 1950s making pointed comment in a certain direction that "academic research was like a sacred cow that had to be fed and watered, but need not yield milk". There certainly are difficulties in translating research results to the shop floor. What is often forgotten is the balance between the expenditure on the original research and the development costs in relation to the time subsequently available to profit from the discovery, before protection ceases and the advance is available to competitors. Thus, focused research cannot be a leisurely occupation. There is no doubt that several potential improvements to practice or products have been 'killed off' for that reason. There may have been a reluctance to integrate research effort in companies with production needs, even where need has been identified. With involvement of production staff from the beginning of a project, there may be more assurance of a willingness to translate results to the plant, even perhaps to the extent of a contract to do so.

I cannot help feeling also, that the relatively poor position of the engineering industry in Britain today in relation to our heritage has something to do with the image and the reward of the engineer and the scientist in our present society. A culture has developed in which the legal, investment and accountancy professions, paying high salaries, are seen as offering superior opportunities for the young, an image strengthened by a knowledge that top management today has often not scientific or engineering credentials. A trainee accountant may be offered considerably more than even a post-doctoral metallurgist seeking a move to industry.

John Percy's closing words in his address at the opening of the Royal School of Mines were: "In proportion to the success with which the metallurgic art is practised in this country, will the interests of the whole population, directly or indirectly, in no inconsiderable degree be promoted". I wonder what he would think about this today.

WHITHER METALLURGY?

Recent decades have seen the virtual elimination of the 'metallurgist' in favour of the 'materials scientist' in academic and institutional circles. The metallurgist's role was to link knowledge (science) to practice (technology) in relation to a particular industrial activity, supporting plant engineers in the processing stages between mineral and finished metal product, often with eventual professional specialisation as regards the particular stage covered (thus extraction or process metallurgists and physical metallurgists) or, even, the metal being exploited. Just as with any engineer in industry it was his/her remit to ensure that financial viability in production was achieved.

Increasingly, there was an interest in applying the principles that had been developed in physical metallurgy, in controlling and improving the properties obtainable, to non-metallics. With the expansion of metallurgy departments in the 1960s and onwards and the increasing emphasis on physical metallurgy and these 'new' materials, the recruitment into the teaching strength of departments which could no longer be of metallurgists, in short supply, was often of physicists, who were naturally often less concerned with the technology of the subject. They were well placed to exploit the science-based expansion into non-metallic materials, bringing a more mathematical and analytical approach to the subject where metallurgy was the seed bed, and extending the studies of properties beyond the chemical and mechanical. At the same time, the pure science departments were increasingly taking a scientific interest in developments in materials generally. Also, engineers were quite properly increasingly introducing materials engineering into their own courses and in some cases metallurgy departments were subsumed by them. Thus, academically, metallurgy was being squeezed from within and without. In what were previously metallurgy courses, expansion into ceramics, polymers and composites meant the curtailment of other topics and 'materials science' departments were born. This was not a development confined to Cambridge but was national, nay international. The expansion of many courses to four years has been used to compensate for some of the losses to metallurgy. When I first taught at Cambridge, the final year course alone had 12 lectures on the science and technology of iron- and steelmaking. Currently the students will be offered a maximum of six lectures on phase equilibria and six on heat and mass transfer in the third year, without any especial attention to a particular thermochemical process, to cover right across the processing spectrum. Twelve lectures on extraction and recycling in general are an option in the fourth year. Whilst the lecturers will clearly use real examples wherever possible, it still may not equip a graduate to play a significant role in say, steelmaking or secondary metal recovery, without subsequent additional training. British Steel are known to provide such further training, having come to terms with the level of knowledge and understanding in their field shown by present day graduates.

It has to be admitted that at the same time, there has been a decline in the extraction and even in the secondary metal recovery activities in this country, which could be reflected in the popularity of these aspects in university circles, but such is the globalisation of industries these days that good metallurgists in the field can find employment, not only here where there are jobs available, but elsewhere in the world. The problem extends, however, to recruitment in relation to engineering firms and foundries where the companies are often small and could not themselves supply postgraduate training. For example, few material science courses provide enough understanding of steel selection or microstructure development in steels to give required properties, and the complexities of heat treatment, for graduates to be able to take a responsible role in, say, a heat treatment firm. To negotiate with customers and supply to specification requires good understanding and there really is a shortage of qualified metallurgists at present in the heat treatment field.

There has been a reduction in the theoretical teaching of solidification and of practical foundry technology, with often no opportunity for undergraduates to actually 'handle' molten metal and to make a casting. No amount of 'modelling' can replace the insight and interest from practical experience. On arriving at Cambridge, I instituted a casting practical similar to the one that I had enjoyed at the RSM, which had been such an advantage to me when employed at J. Stone. As well as making a step-bar casting in modified aluminium–silicon alloys, illustrating the significance of section and cooling rate and the requirements for microstructural modification, students were able to cast a 'trophy' in the form of a shield or ash-tray bearing the university crest. After equipping the laboratory with foundry sand, moulding boxes, moulding tools and a small melting unit I initially demonstrated the making of moulds and the practice for melting and modification. The practical was popular with the students and I am delighted that at Cambridge an updated version is still taught by Dr E.R. Wallach (Rob), to include the effect of cooling rate on mechanical properties as well as microstructure. The sodium level is determined by sodium sensors of the type developed by Derek Fray in the Department. As another example of where some courses are now failing, although development in alloy cast irons has been one of the most significant in metallurgy in recent years, most materials science students will not even be aware of the complexities of normal cast and ductile irons.

Even more importantly there has been a steady 'drip decline' in the teaching of physical metallurgy in general and metallography. Although metallography, in the context of advanced physical methods, has moved forward enormously in the last 50 years, I doubt that the amount of teaching now given on the interpretation of optical microstructures would merit the four day metallography practical examinations that were in place at the RSM in 1947, and yet such a form of investigation is still the first line of examination in relation to metal failures in the field. In fact, sadly, most courses do not now have a practical examination of any sort. Metallography was one of the important factors that determined that I would not switch to Mining Geology during my undergraduate course and I believe that it has always remained a key factor in making metallurgy courses attractive to students.

Dynamic change is usually a good thing and, no doubt, metallurgy had to move forward, but the broadening of courses to encompass so much on other materials has meant that many features, that I still regard as essential to a metal-lurgist, have been lost. No doubt a ceramicist or other materials expert would feel the same about his particular field. Maybe we have thrown lots of babies out with the bathwater! Surely there have to be places where detailed metallurgy courses can continue to be taught.

Over the years, there has been a gradual change in the name of the Cambridge Department from Metallurgy to Metallurgy and Materials Science and then to Materials Science and Metallurgy. It was often jokingly said by other members of staff that when JAC *finally* retired Metallurgy would be dropped from the title. Perhaps that time is now not far away!

FINALE

That is my 'professional' story. I must end. Yet, I find it singularly difficult to do so, for I have found much pleasure in the unfolding. Whilst memory may be fickle, and I apologise for almost inevitable errors, one memory from the rich-ness of that store has led to scores of others crowding in, some happy, some less so. Some mistakes, difficulties, hopefully a few successes and certainly many characters, the latter standing out more vividly in the recalling.

Although it has not at any time made me a wealthy man, I count myself ex-tremely fortunate to have had a profession that I have enjoyed. In an obituary for my father, killed in an accident when I was a small boy, the following was writ-ten: "His work was not merely a daily toil, but an absorbing interest". So it has been for me.

ACKNOWLEDGEMENTS

A good Secretary is to be treasured! At Stones and BOC, secretarial assistance was given by a Group Secretary or by a small pool. On arriving at Cambridge I was allotted a major part of the time of Pat Evans, who, it turned out, was mar-ried to an old school friend of mine, Tony Evans, and an excellent relationship ensued with both. After she left, there was again a period without a particular nominated secretarial resource until the arrival of the excellent Heather Thomson in 1978. When she left to follow her RAF husband, she was succeeded by Phylis Summerfield who continued to look after Gerry Smith and myself very well for several years until her, and our, formal retirement. We now occasionally add to the load carried by the Head of Department's secretary, Jane Temple or by

Christine Carey who never turn us away and are kindness personified. I am also indebted to Chris Cross, John Langham, David Melford, Ken Sargent, Alfred Webb and David West for helpful comments and to Brian Barber and Carol Best for photographic assistance. My thanks also go to Miss J.S.M. Dannatt and to Mrs W.J.B. Chater for providing the photographs of Professor Dannatt and Bill Chater respectively.

In the preparation of this text I have relied on Mrs Susan Mansfield at St John's College. She has been wonderful and I thank her and the College for their support. I am grateful, also, to the Books Committee of The Institute of Materials and the head of the Books Department, Peter Danckwerts, for encouraging me in this venture, and to Miss Jo Jacomb for the production work.

Index